Spatial Problem Solving

with Cuisenaire® Rods

by

Patricia S. Davidson
Robert E. Willcutt

Cuisenaire Company of America
12 Church St. • Box D
New Rochelle, NY 10805

A Special Note to Teachers

This workbook has been written with the intention of meeting some of the curricular needs uncovered by recent neurological research, specifically the need to make a more conscious effort to develop right hemispheric skills. Neurological research indicates that the two sides of the brain perform distinct functions: the left hemisphere being critical to verbal, logical, and sequential processing; and the right hemisphere specializing in spatial, perceptual, and holistic processing. It has been asserted that our culture and especially our school systems have overly stressed left hemispheric skills, often to the detriment of people's spatial reasoning powers.

This workbook provides valuable spatial problem solving experience of three major types: space-filling activities with rods similar in spirit to Tangram puzzles; rotation and reflection problems presented separately and together in analogy puzzles and sequential chains; and architectual-type activities involving two-dimensional drawings (top views, front views, and side views) of three-dimensional rod designs. The students need to use the rods in solving all the problems so that the concepts will have concrete foundation.

The activities can be done individually, in small groups, or with a whole class. Students should be encouraged to compare and discuss their answers and problem solving strategies. Each of the worksheets provides working space for the rods and spaces for the answers for ease in implementation. Teaching suggestions and answers are given on Pages 52-58. A centimeter graph paper master and master cut-outs for use with specific pages are provided on Pages 59 and 60.

Even though the workbook is intended primarily for students in Grades 4-8, adults of all ages will also benefit from these challenges to their spatial prowess.

Copyright © 1983 by
Cuisenaire Company of America, Inc.
12 Church Street, New Rochelle, NY 10805

ISBN 0-914040-99-5

The name Cuisenaire® and the color sequence of the rods
are trademarks of the Cuisenaire Company of America, Inc.

Permission is granted for limited reproduction of pages from this book for classroom use.

Printed in U.S.A.

TABLE OF CONTENTS

Rod Tangram-Type Problems
    Covering Designs with One Rod of Each Color   1-4
    Covering Designs with a Given Set of Rods   5-8
    Covering Designs with a Specific Number of Rods   9-14

Rotations
    Rotating Designs in a Clockwise Direction   15
    Drawing Rotated Rod Designs   16
    Rotating Different Rod Designs   17-18
    Finding the Type of Rotation   19-20

Reflections
    Reflecting Rod Designs   21
    Drawing Reflected Rod Designs   22-23
    Reflecting Different Rod Designs   24
    Finding the Type of Reflection   25-26

Rotations and Reflections
    Learning How to Solve Rod Analogy Puzzles   27
    Solving Rotation and Reflection Analogy Puzzles   28-34
    Reversing Rod Analogy Puzzles   35-36
    Solving a Chain of Rotations and Reflections   37-40

Drawings of Three-Dimensional Rod Designs
    Seeing Three-Dimensional Rod Designs in Two Dimensions   41
    Drawing Three-Dimensional Rod Designs in Two Dimensions   42-43
    Drawing the Three Views of a Three-Dimensional Rod Design   44-46
    Identifying Rod Designs from Two-Dimensional Drawings   47-48
    Building Rod Designs from Two-Dimensional Drawings   49-51

Selected Answers and Comments   52-58

Centimeter Graph Paper Master   59

Master Cut-out Sheet   60

COVERING DESIGNS WITH ONE ROD OF EACH COLOR

Use exactly one rod of each of the ten colors to cover this design. Trace your solution and record the color names to show how you placed the rods. There is more than one correct way.

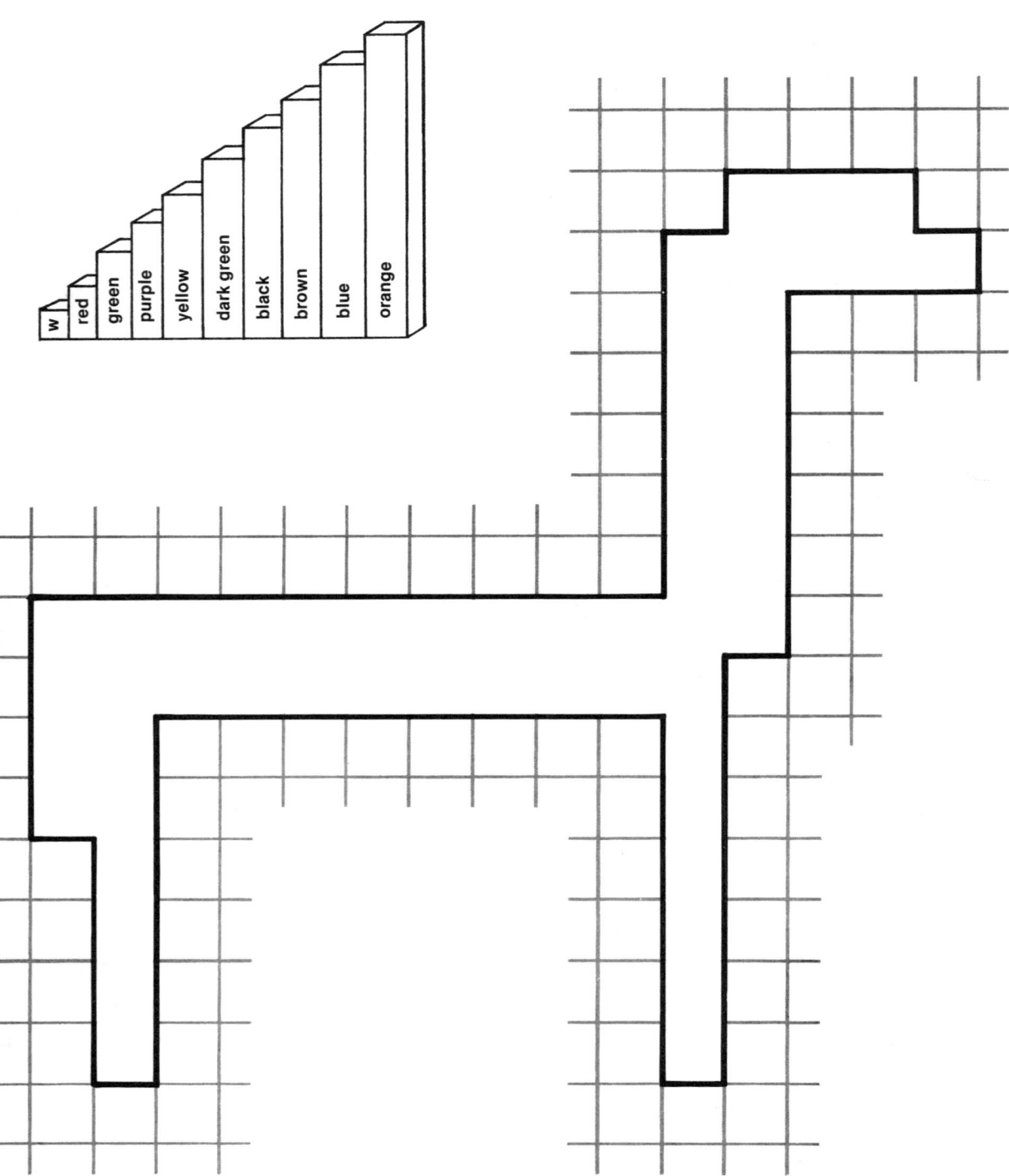

Page 1　　SPATIAL PROBLEM SOLVING with Cuisenaire Rods © 1983 Cuisenaire Co. of America, Inc.

COVERING DESIGNS WITH ONE ROD OF EACH COLOR

Use exactly one rod of each of the ten colors to cover this design. Trace your solution and record the color names to show how you placed the rods. There is more than one correct way.

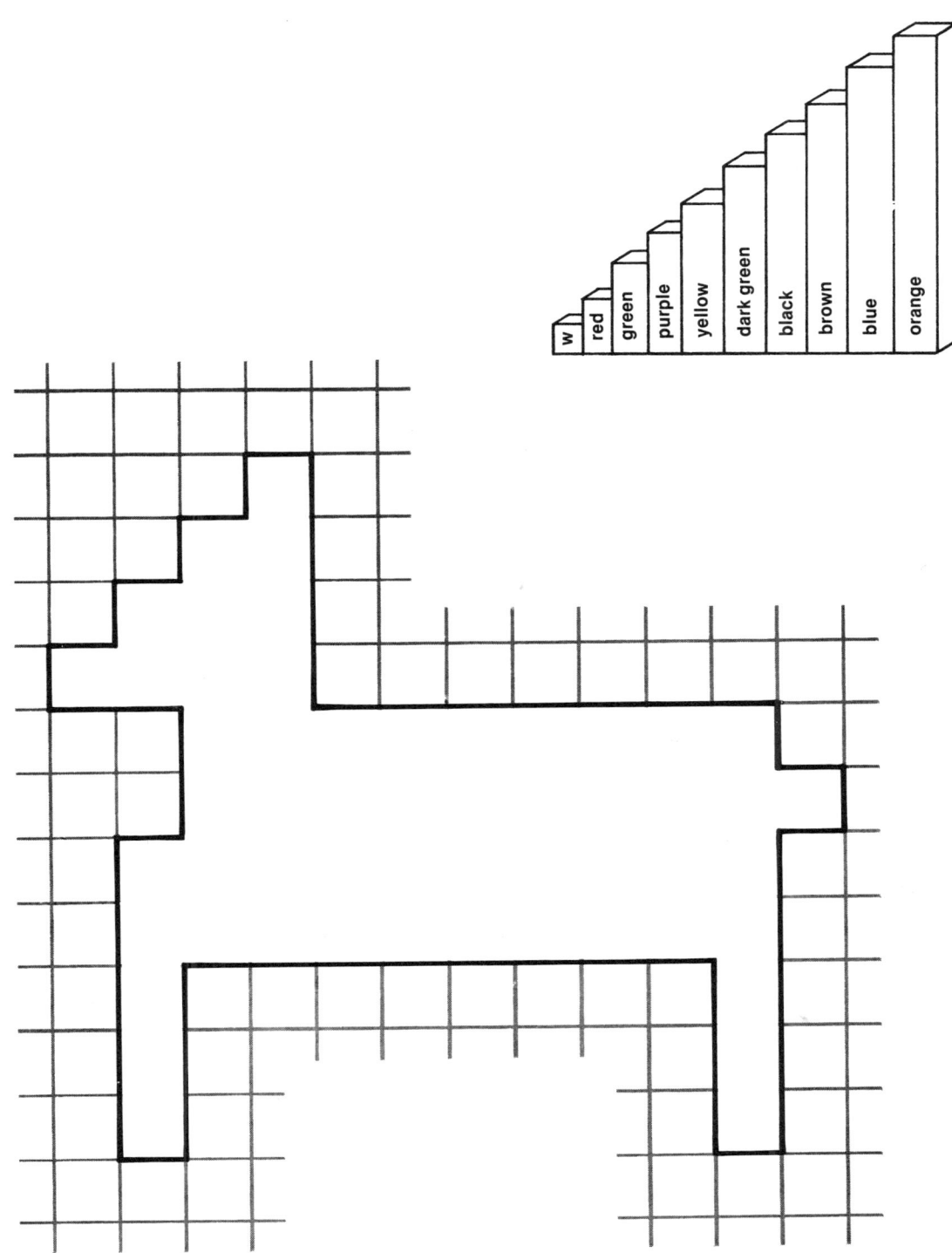

Spatial Problem Solving with Cuisenaire Rods © 1983 Cuisenaire Co. of America, Inc.

COVERING DESIGNS WITH ONE ROD OF EACH COLOR

Use exactly one rod of each of the ten colors to cover each design. Trace your solution and record the color names to show how you placed the rods.

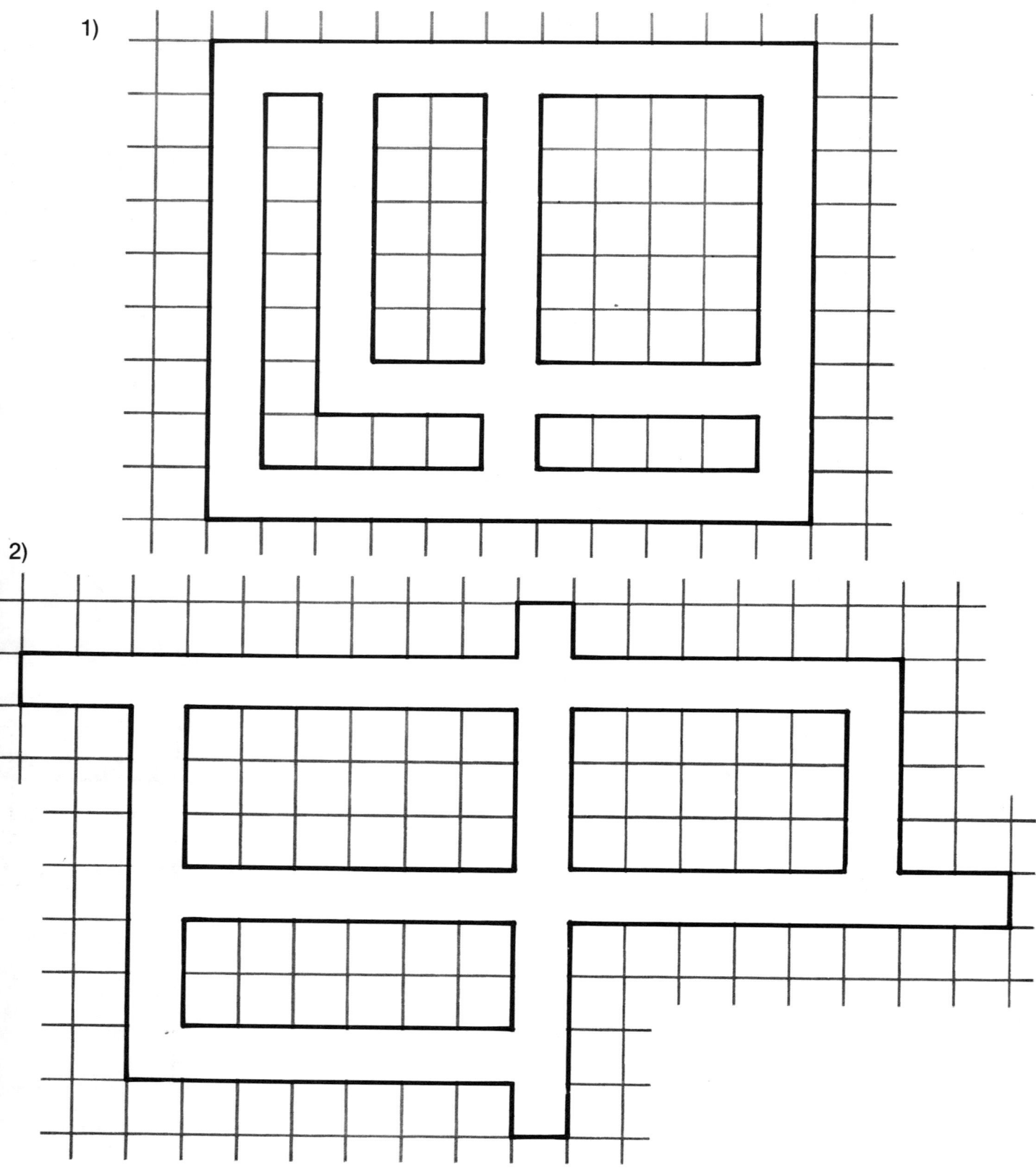

Page 3   SPATIAL PROBLEM SOLVING with Cuisenaire Rods © 1983 Cuisenaire Co. of America, Inc.

COVERING DESIGNS WITH ONE ROD OF EACH COLOR

Use exactly one rod of each of the ten colors to cover each design. Trace your solutions and record the color names to show how you placed the rods.

1)

2)

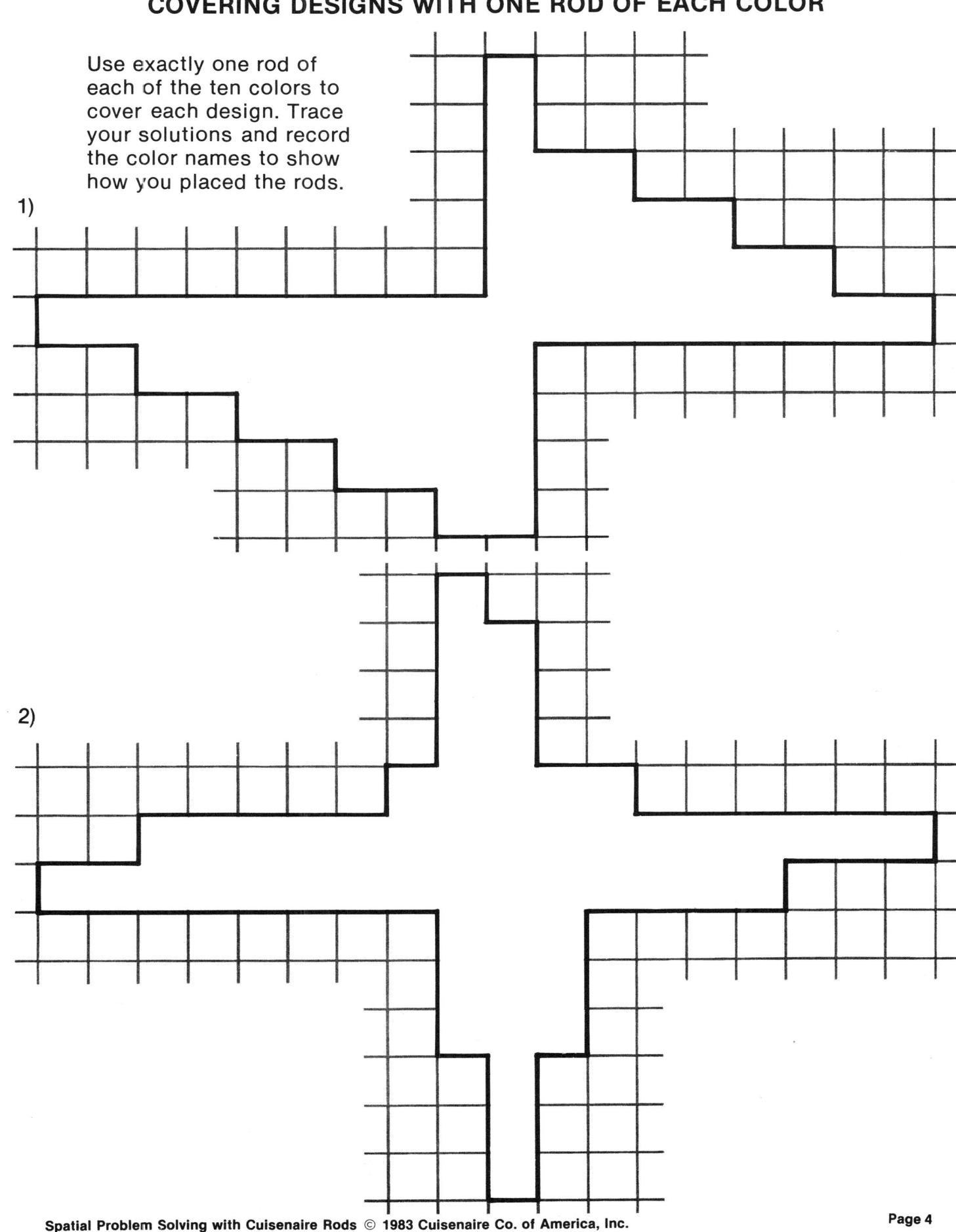

SPATIAL Problem Solving with Cuisenaire Rods © 1983 Cuisenaire Co. of America, Inc.

Page 4

COVERING DESIGNS WITH A GIVEN SET OF RODS

Use the given set of rods to cover each design. Record the color names on the design to show your solution.

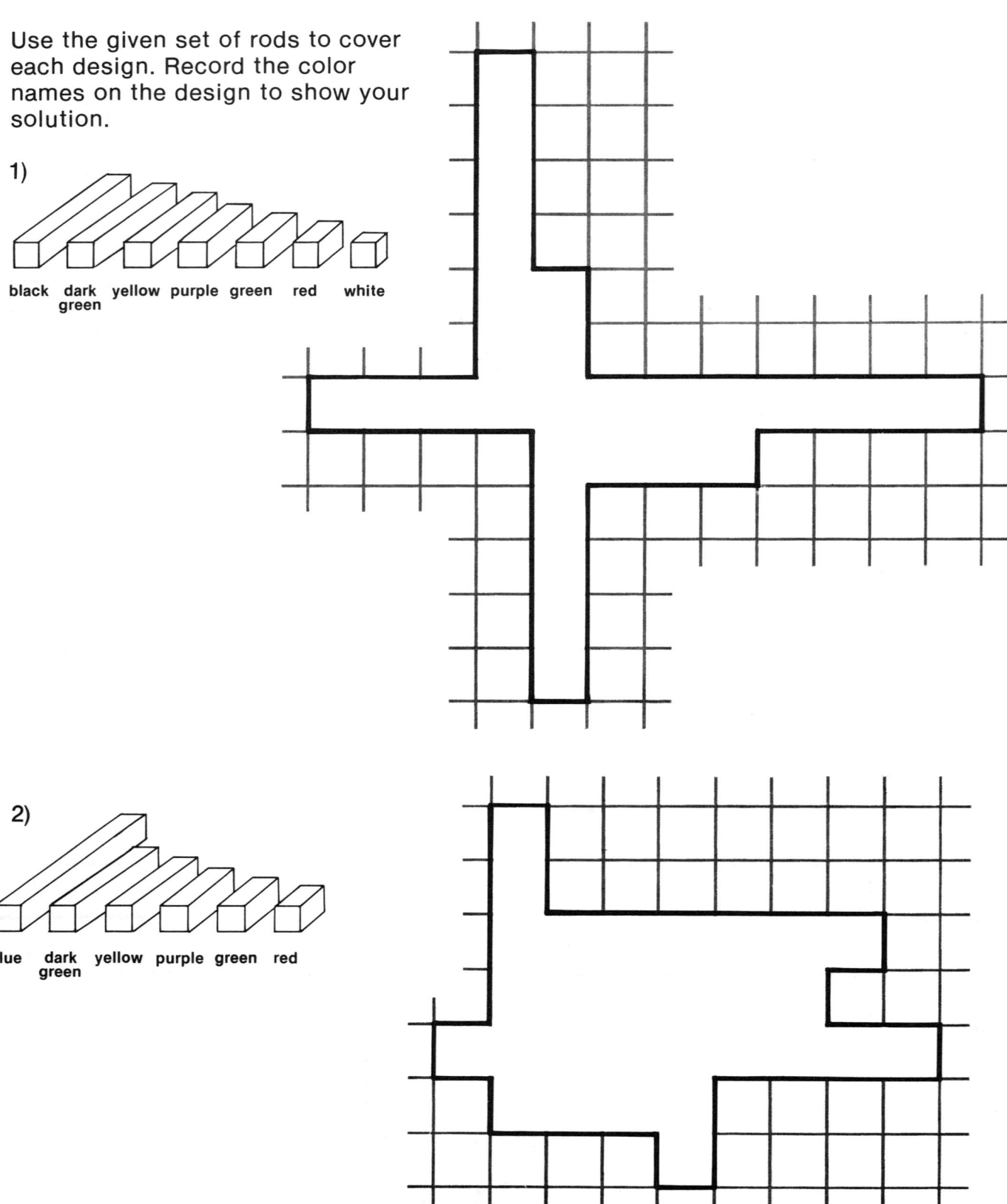

1) black, dark green, yellow, purple, green, red, white

2) blue, dark green, yellow, purple, green, red

Using the given rods, can these be done in more than one way?

COVERING DESIGNS WITH A GIVEN SET OF RODS

Use the given rods to cover these designs. Record the color names on the designs to show your solutions. Can these be done in more than one way?

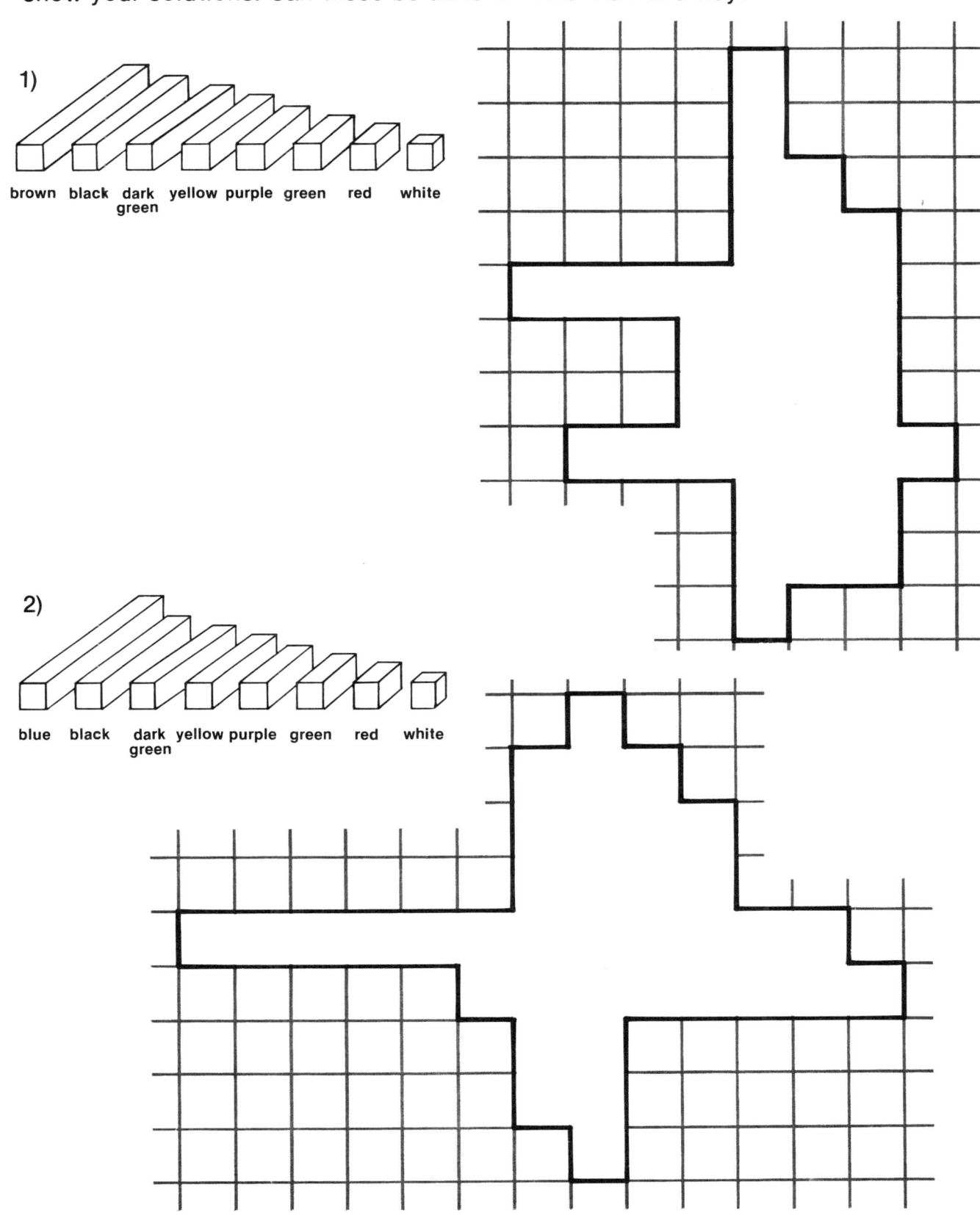

1) brown black dark green yellow purple green red white

2) blue black dark green yellow purple green red white

COVERING DESIGNS WITH A GIVEN SET OF RODS

Use the given rods to cover these designs. Record the color names on the designs to show your solutions. Can these be done in more than one way?

1)

orange blue black dark green yellow purple purple

green green red red white

2)

blue brown black black dark green dark green

yellow yellow purple purple purple

purple green

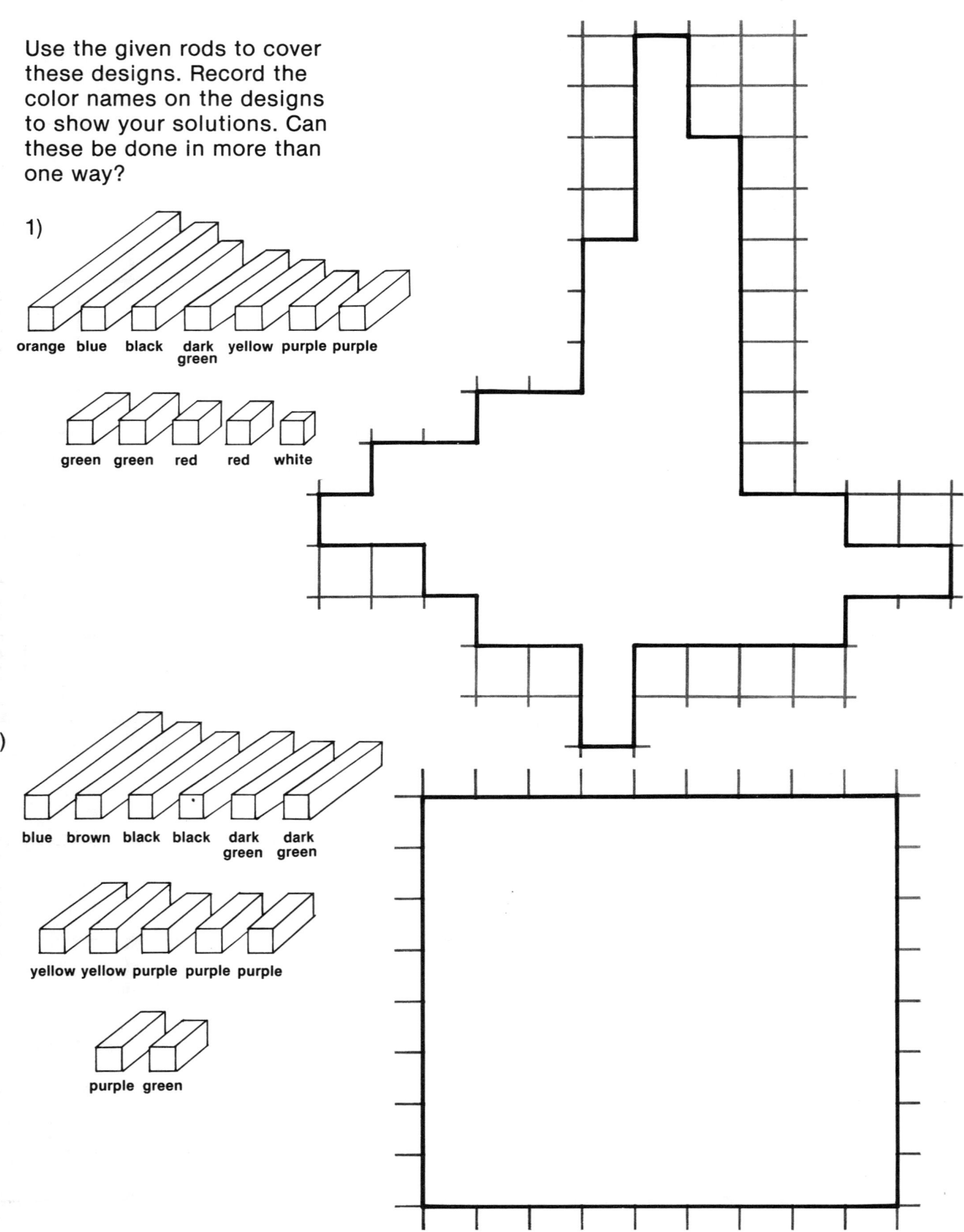

Page 7 Spatial Problem Solving with Cuisenaire Rods © 1983 Cuisenaire Co. of America, Inc.

COVERING DESIGNS WITH A GIVEN SET OF RODS

Use the given rods to cover these designs. Record the color names on the designs to show your solutions. Can these be done in more than one way?

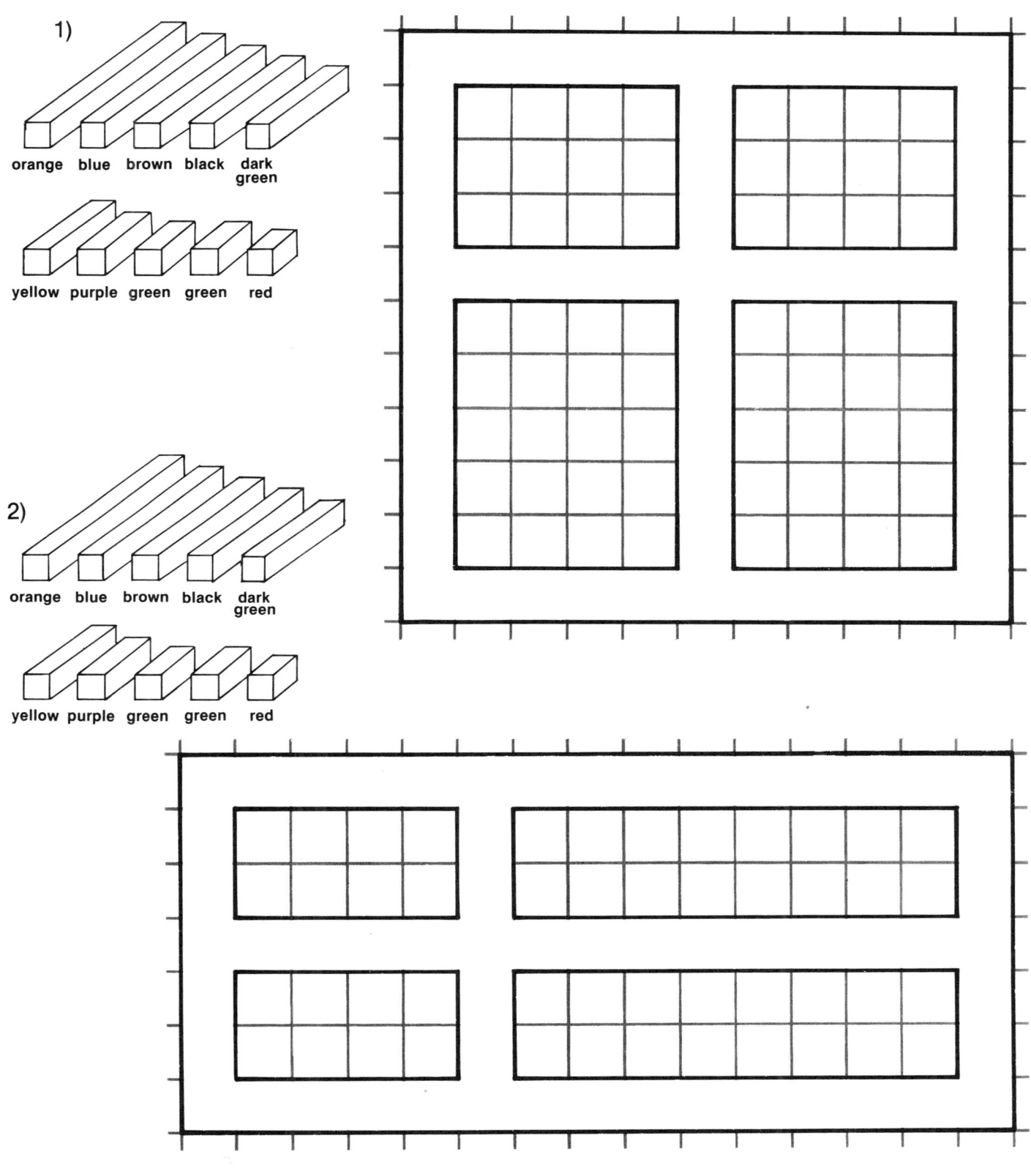

COVERING DESIGNS WITH A SPECIFIC NUMBER OF RODS

1) Use exactly six rods to cover this design. Each rod must be a different color. Record the color names on the design to show your solution.

2) Use exactly seven rods to cover this design. Each rod must be a different color. Record your solution.

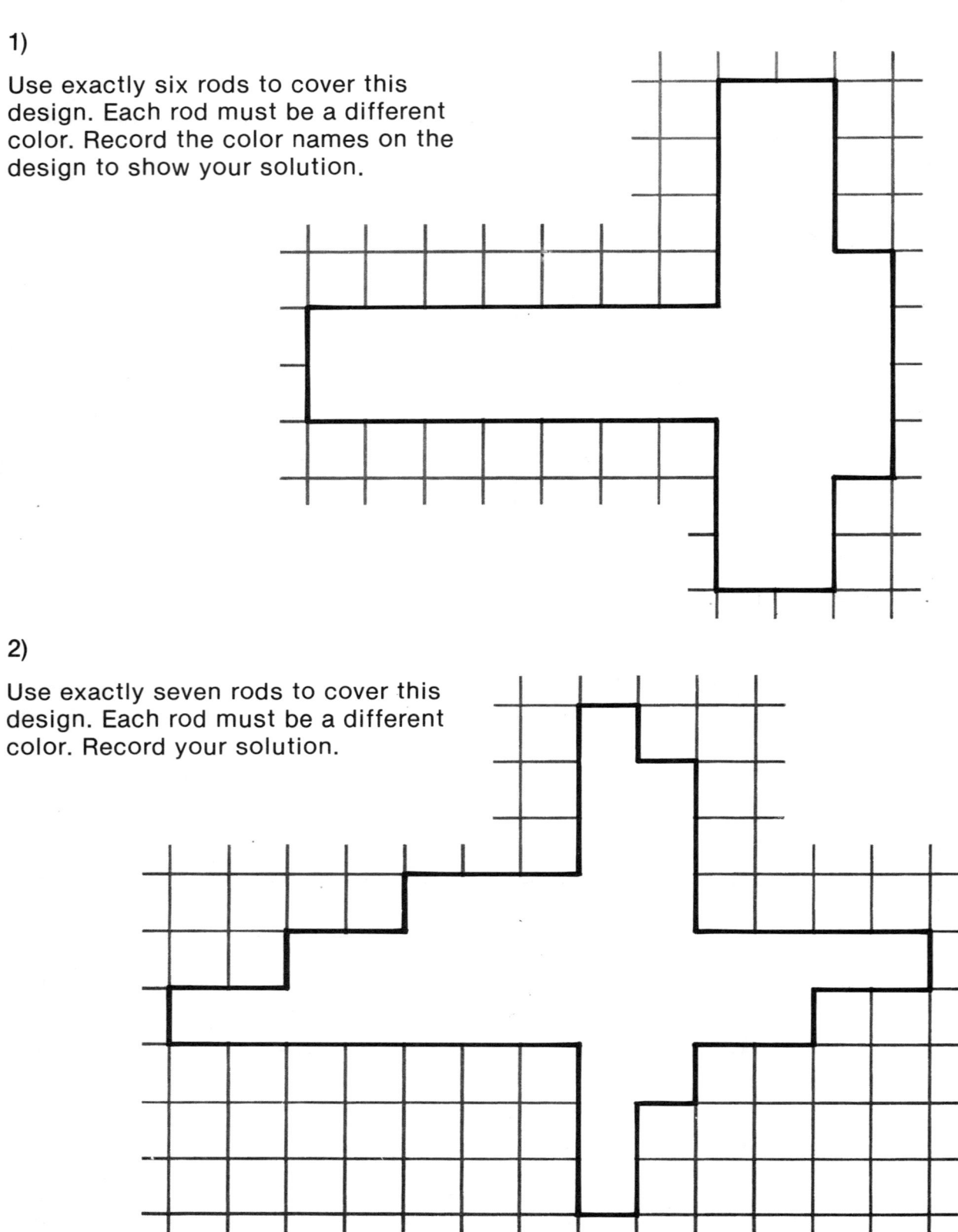

COVERING DESIGNS WITH A SPECIFIC NUMBER OF RODS

1) Use exactly seven rods to cover this design. Each rod must be a different color. Record the color names on the design to show your solution.

2) Use exactly eight rods to cover this design. Each rod must be a different color. Record your solution.

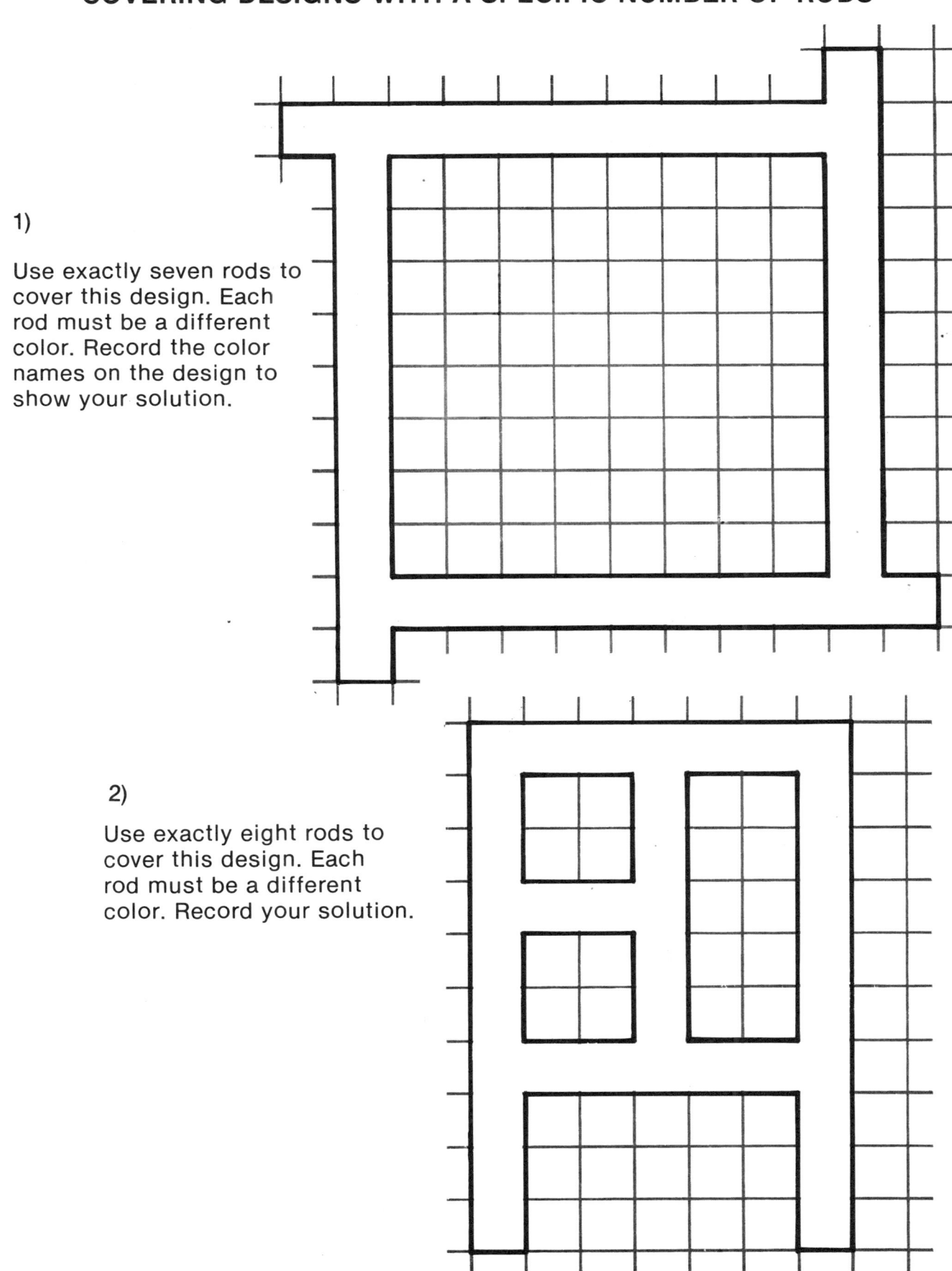

COVERING DESIGNS WITH A SPECIFIC NUMBER OF RODS

1)

Use exactly eight rods and only two colors to cover this design. Record the color names on the design to show your solution.

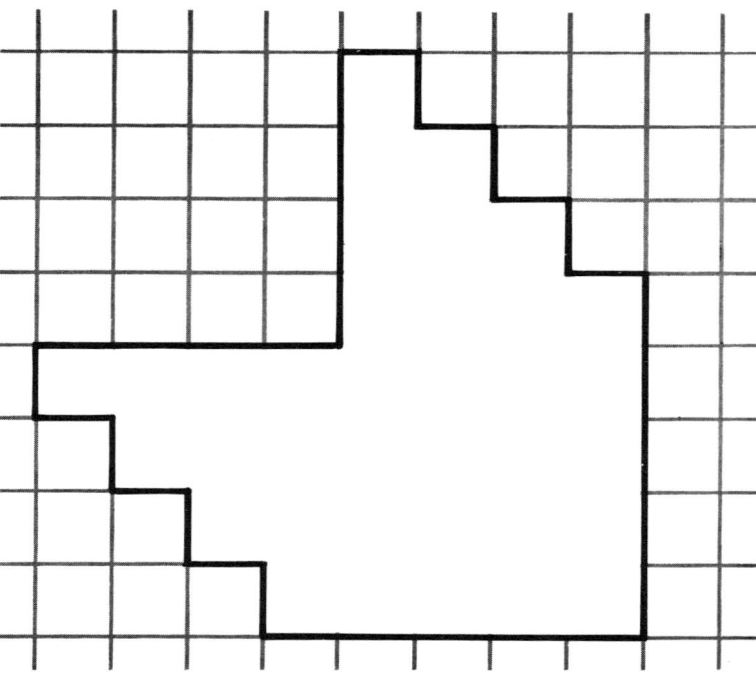

2)

Use exactly twelve rods and only three colors to cover this design. Record the color names on the design to show your solution.

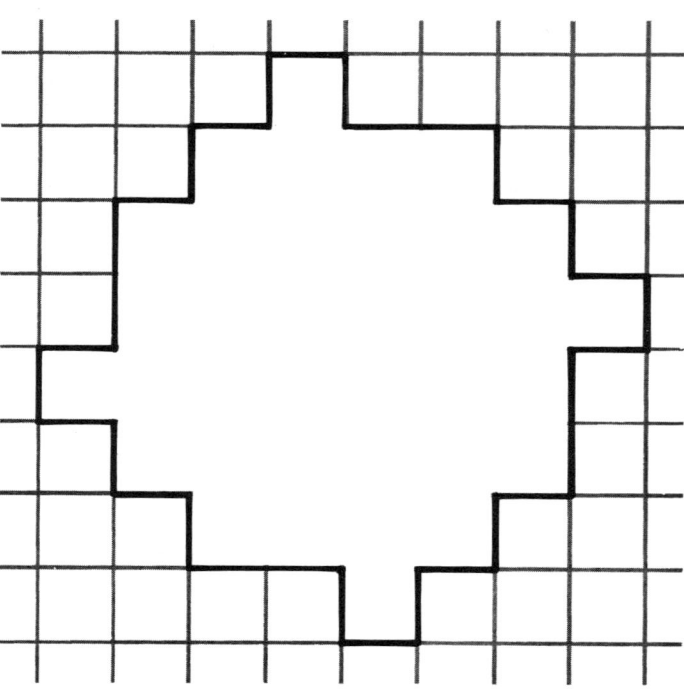

COVERING DESIGNS WITH A SPECIFIC NUMBER OF RODS

1)

Use exactly eight rods and only two colors to cover this design. Record the color names on the design to show your solution.

2)

Use exactly twelve rods and only three colors to cover this design. Record your solution.

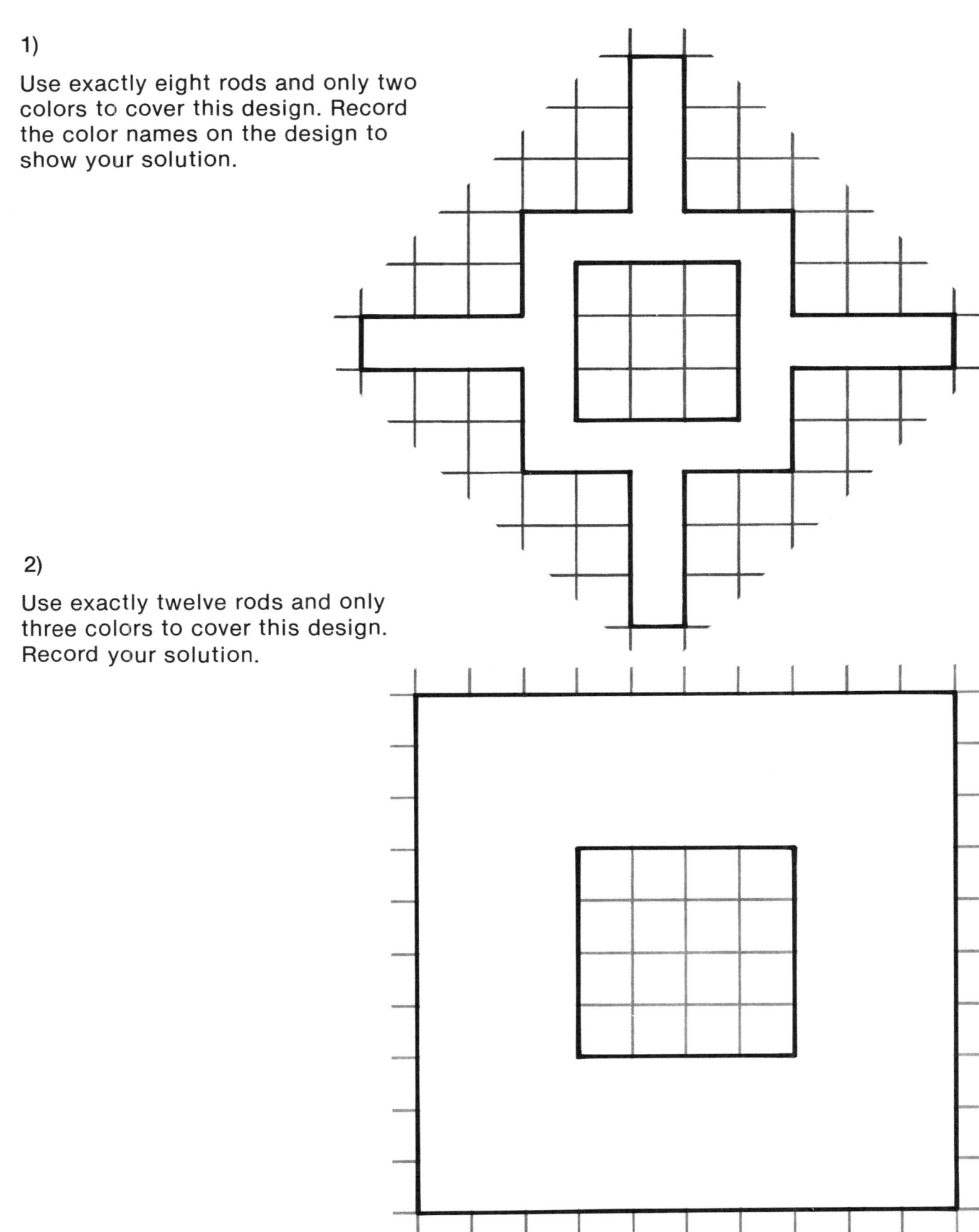

Spatial Problem Solving with Cuisenaire Rods © 1983 Cuisenaire Co. of America, Inc.

COVERING DESIGNS WITH A SPECIFIC NUMBER OF RODS

1) Use exactly twelve rods and only three colors to cover this design. Record the color names on the design to show your solution.

2) Use exactly ten rods and only two colors to cover this design. Record your solution.

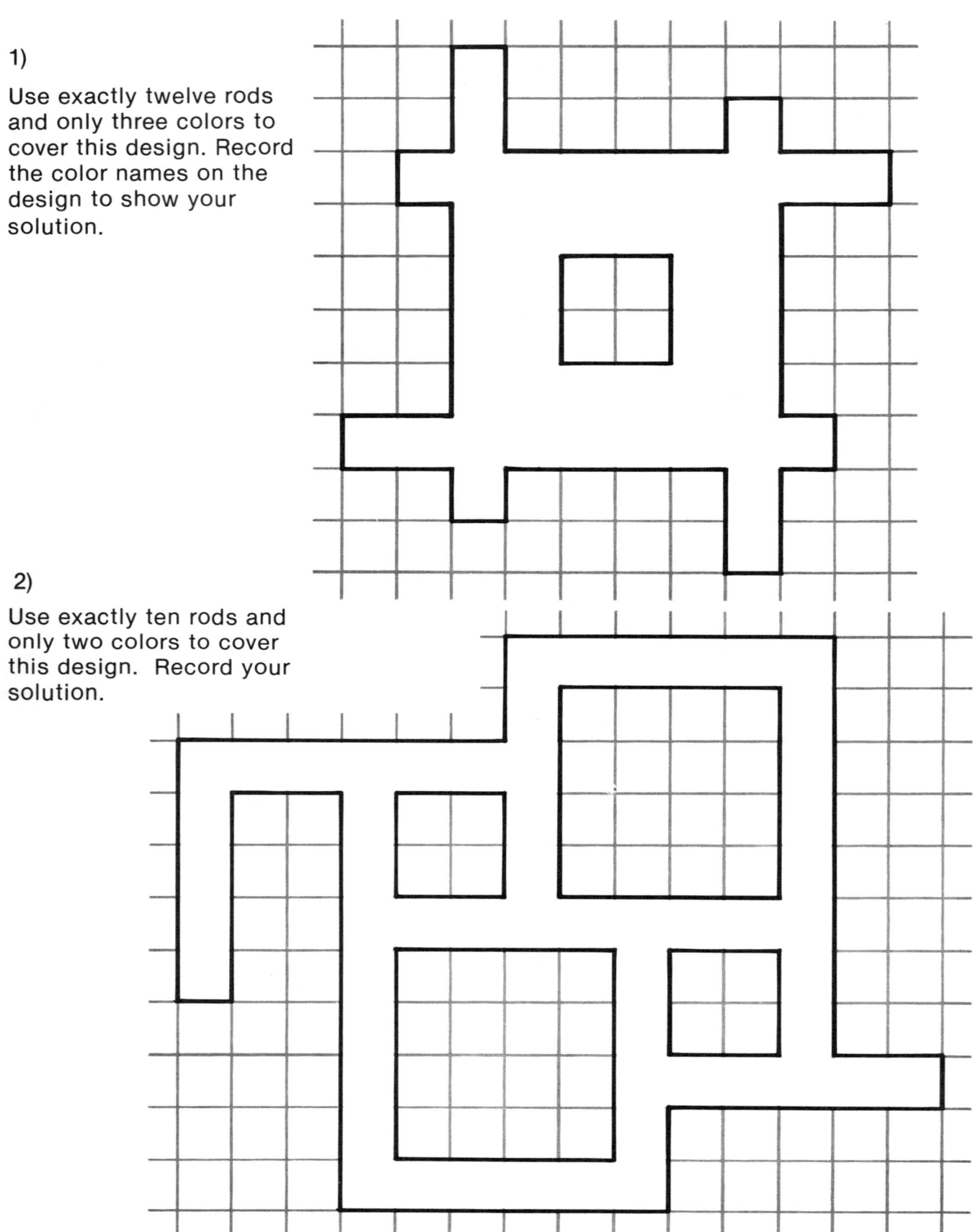

Spatial Problem Solving with Cuisenaire Rods © 1983 Cuisenaire Co. of America, Inc.

COVERING DESIGNS WITH A SPECIFIC NUMBER OF RODS

1)

Use exactly ten rods and only two colors to cover this design. Record the color names on the design to show your solution.

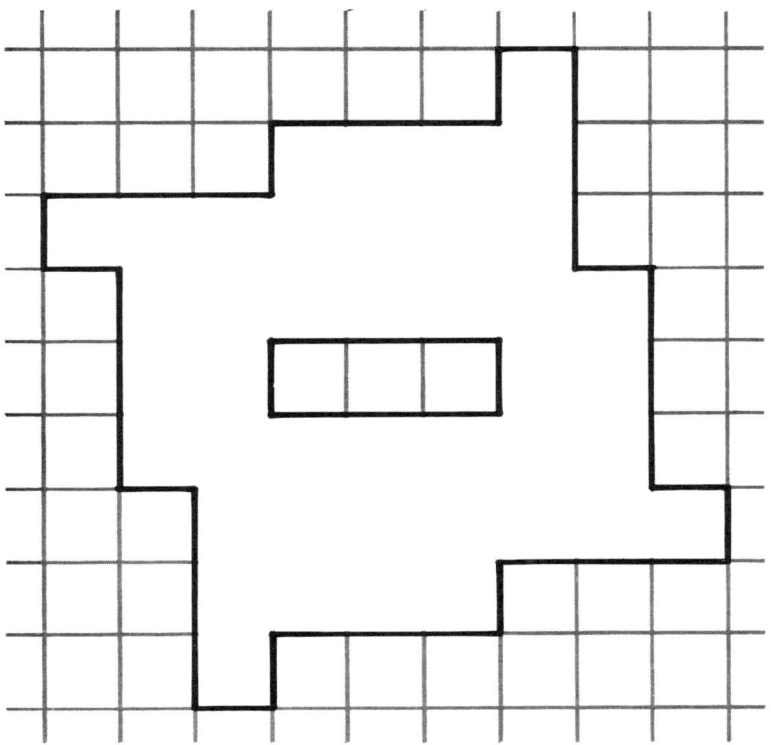

2)

Use exactly fourteen rods and only two colors to cover this design. Record your solution.

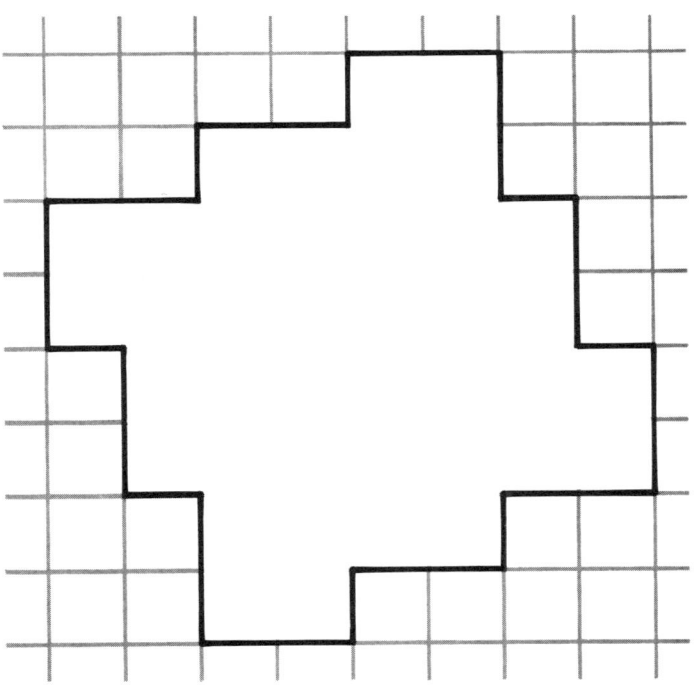

ROTATING DESIGNS IN A CLOCKWISE DIRECTION

This page illustrates what a design looks like after it has been rotated a given number of degrees in a clockwise direction. For each example, make the design as indicated to produce the resulting design in the right column. Compare the original design and the rotated design so that you can visualize the effect of each type of rotation. You may wish to use the Master Cut-Out #1 from page 60 to help you.

Example of a 90° Clockwise Rotation

Example of a 180° Clockwise Rotation

Example of a 270° Clockwise Rotation

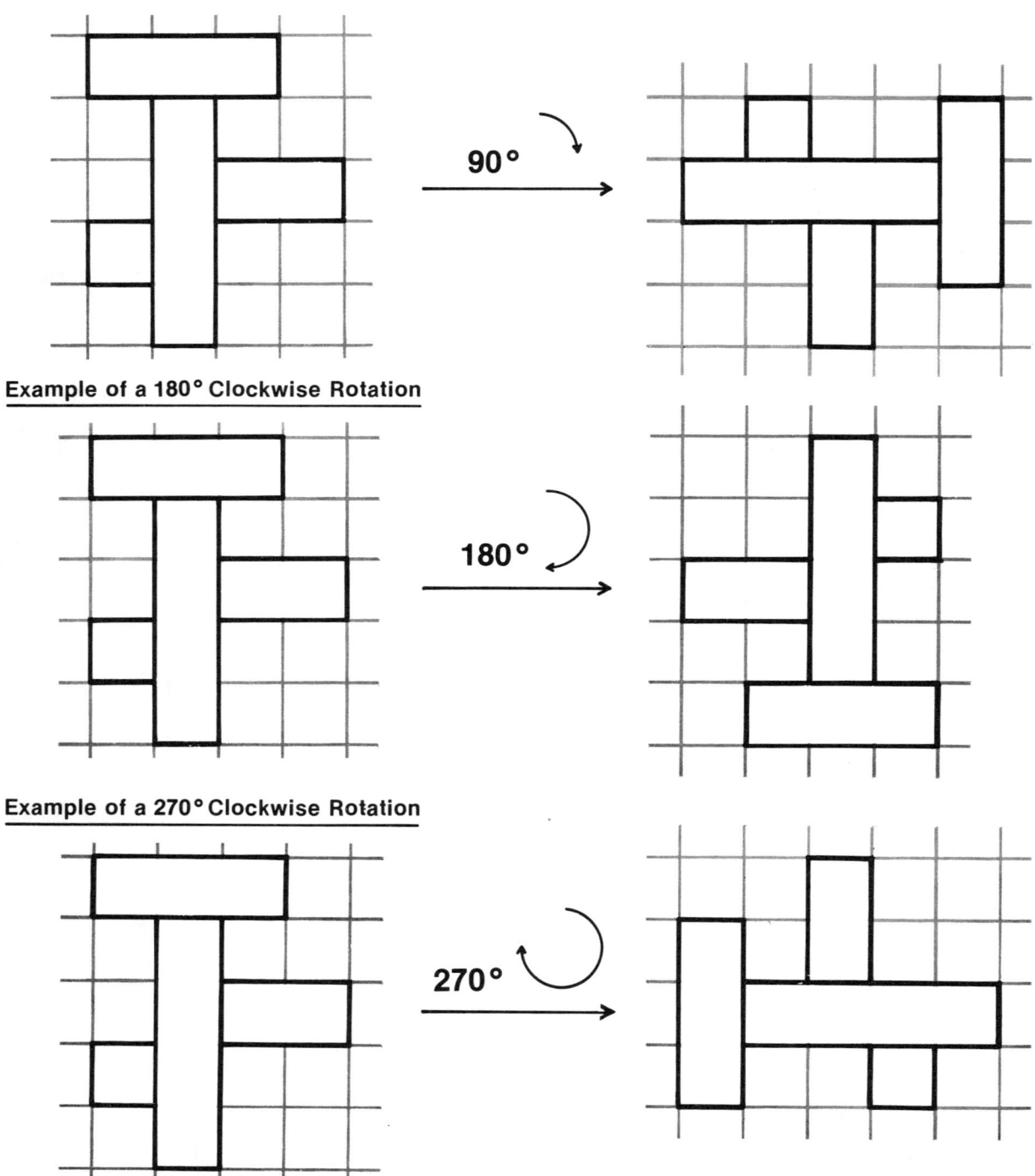

DRAWING ROTATED ROD DESIGNS

Make the given design with rods. Rotate the design in a clockwise direction the number of degrees indicated. Then draw the design resulting from each rotation. Compare the original design and the rotated design so that you can visualize the effect of each type of rotation. You may wish to use the Master Cut-Out #2 from page 60 to help you.

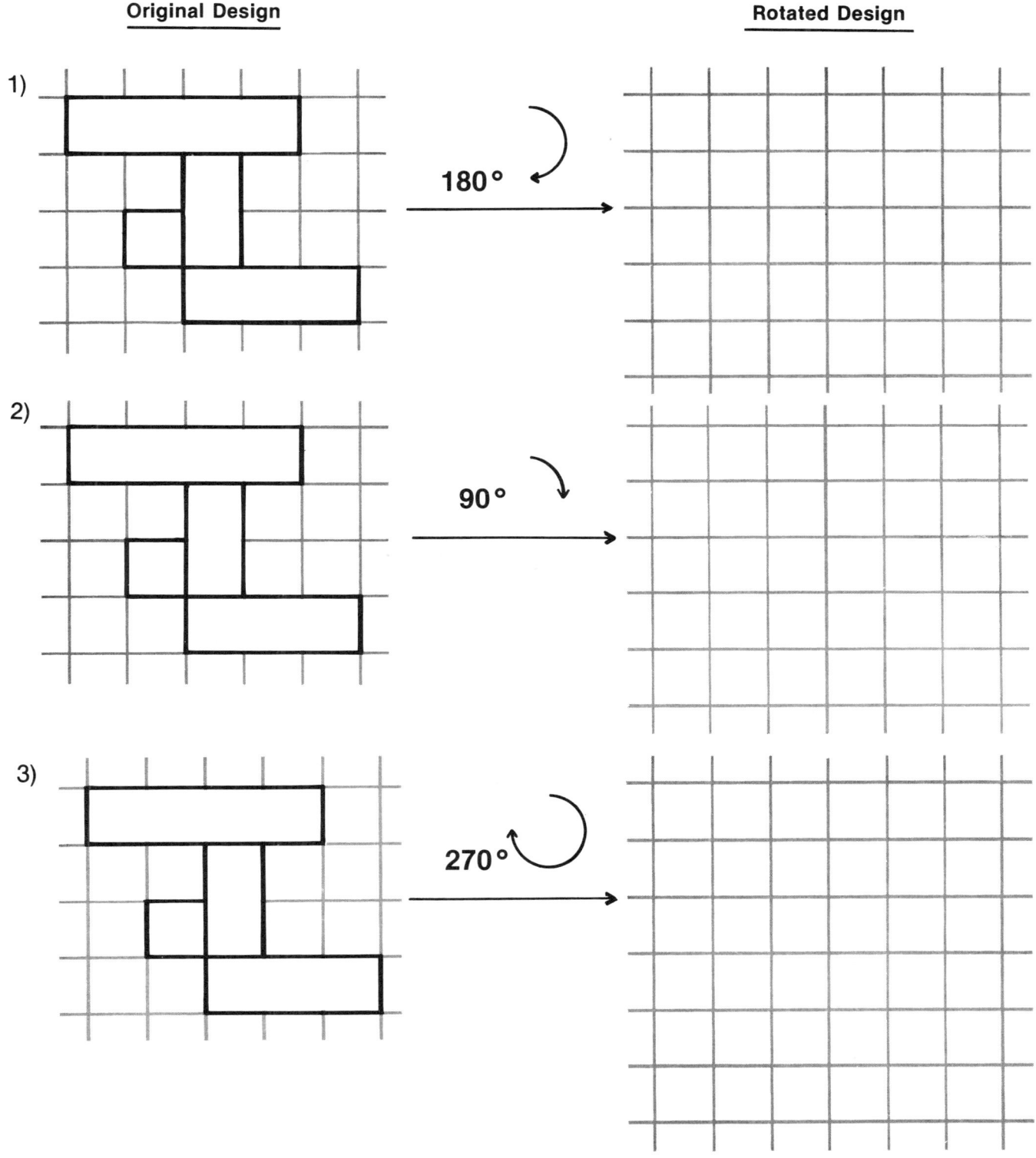

ROTATING DIFFERENT ROD DESIGNS

Make each of the given designs with rods. Rotate the design in a clockwise direction the number of degrees indicated. Then draw the design resulting from each rotation.

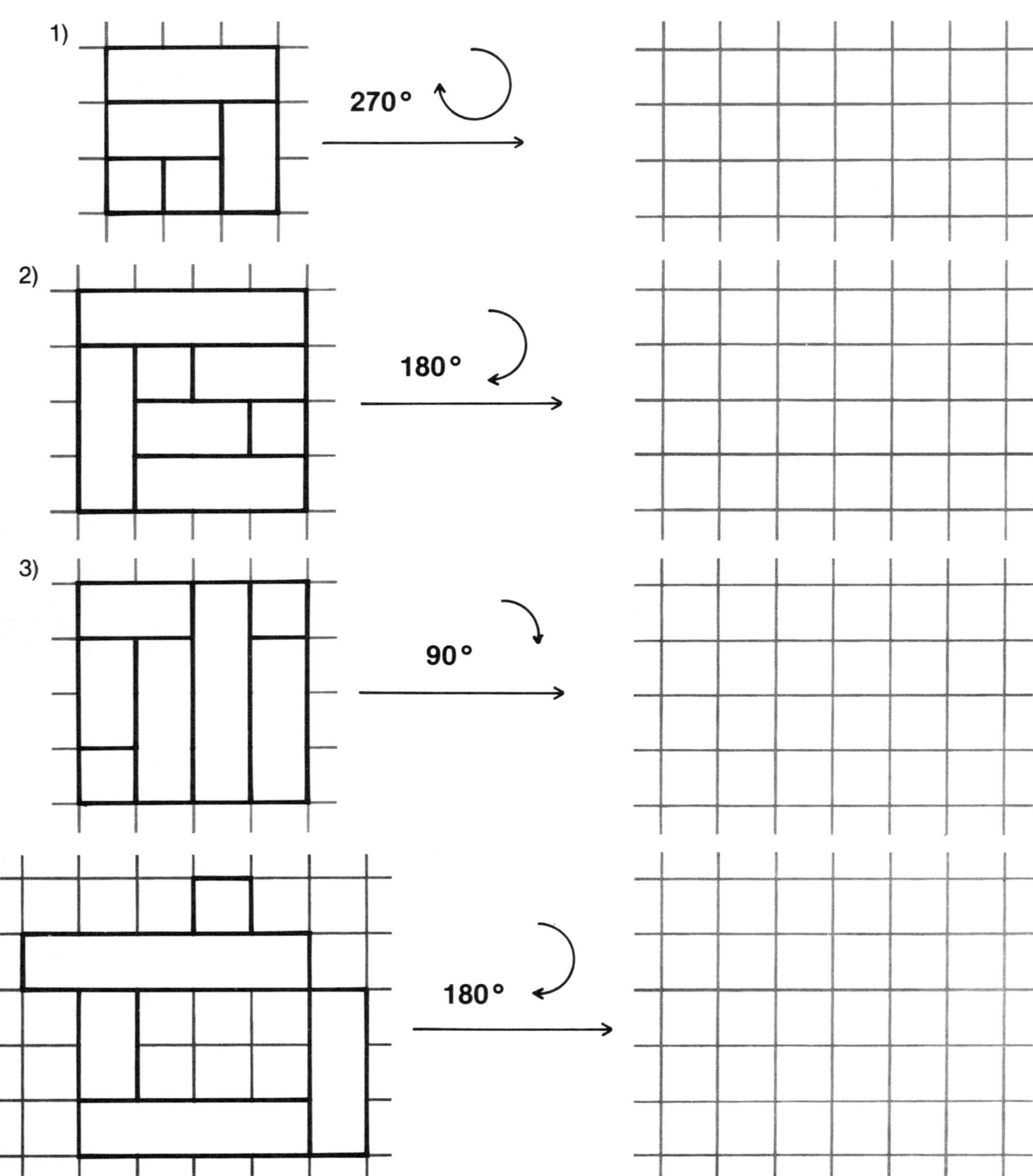

Page 17　　　Spatial Problem Solving with Cuisenaire Rods © 1983 Cuisenaire Co. of America, Inc.

ROTATING DIFFERENT ROD DESIGNS

Make each of the given designs with rods. Rotate the design in a clockwise direction the number of degrees indicated. Then draw the design resulting from each rotation.

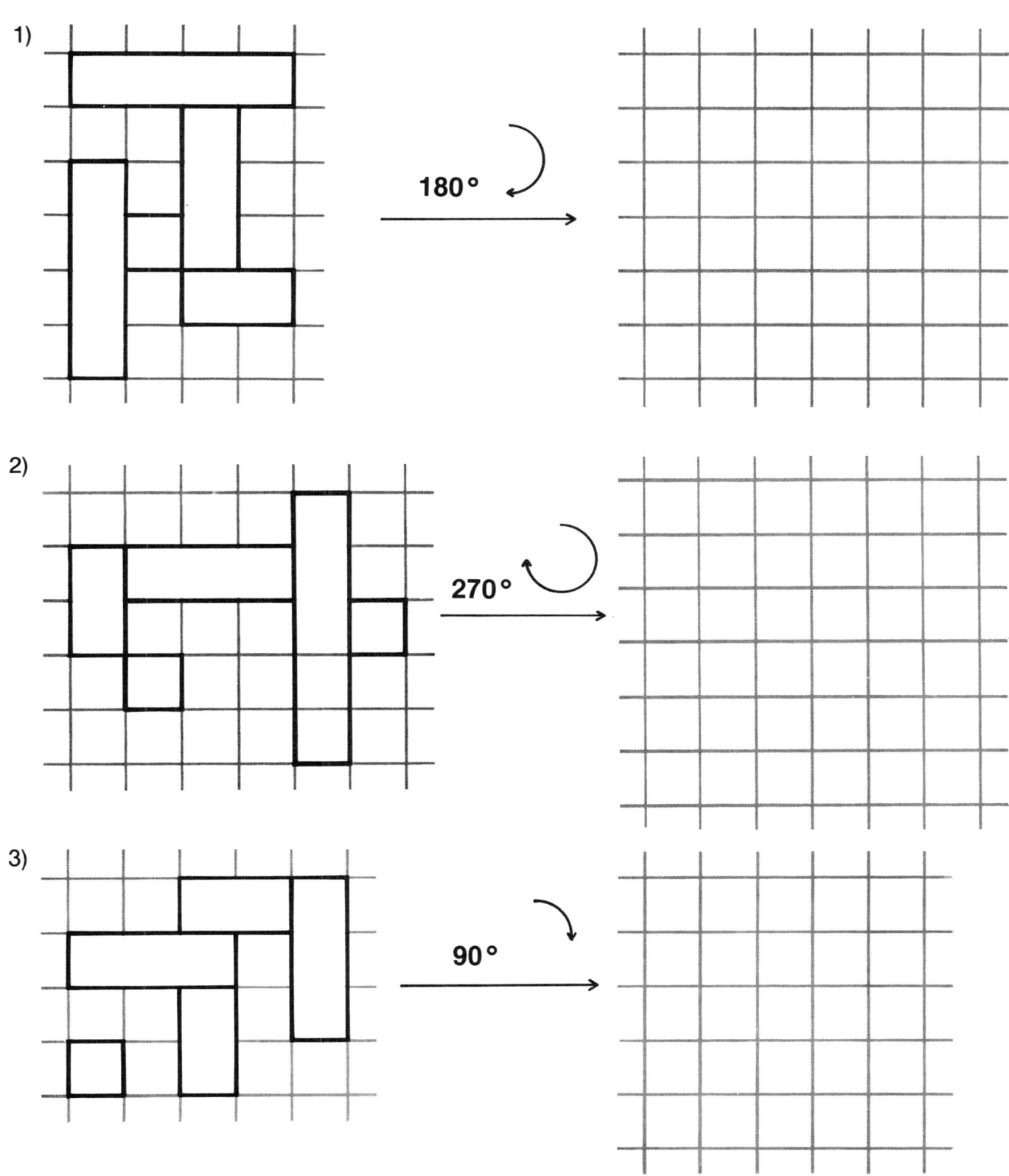

Spatial Problem Solving with Cuisenaire Rods © 1983 Cuisenaire Co. of America, Inc.

FINDING THE TYPE OF ROTATION

Use rods to make the design shown in the left column. Then name the type of rotation that would produce the resulting design shown in the right column.

Fill in 90° ↷, 180° ↶, or 270° ↺ in the circle above the arrow.

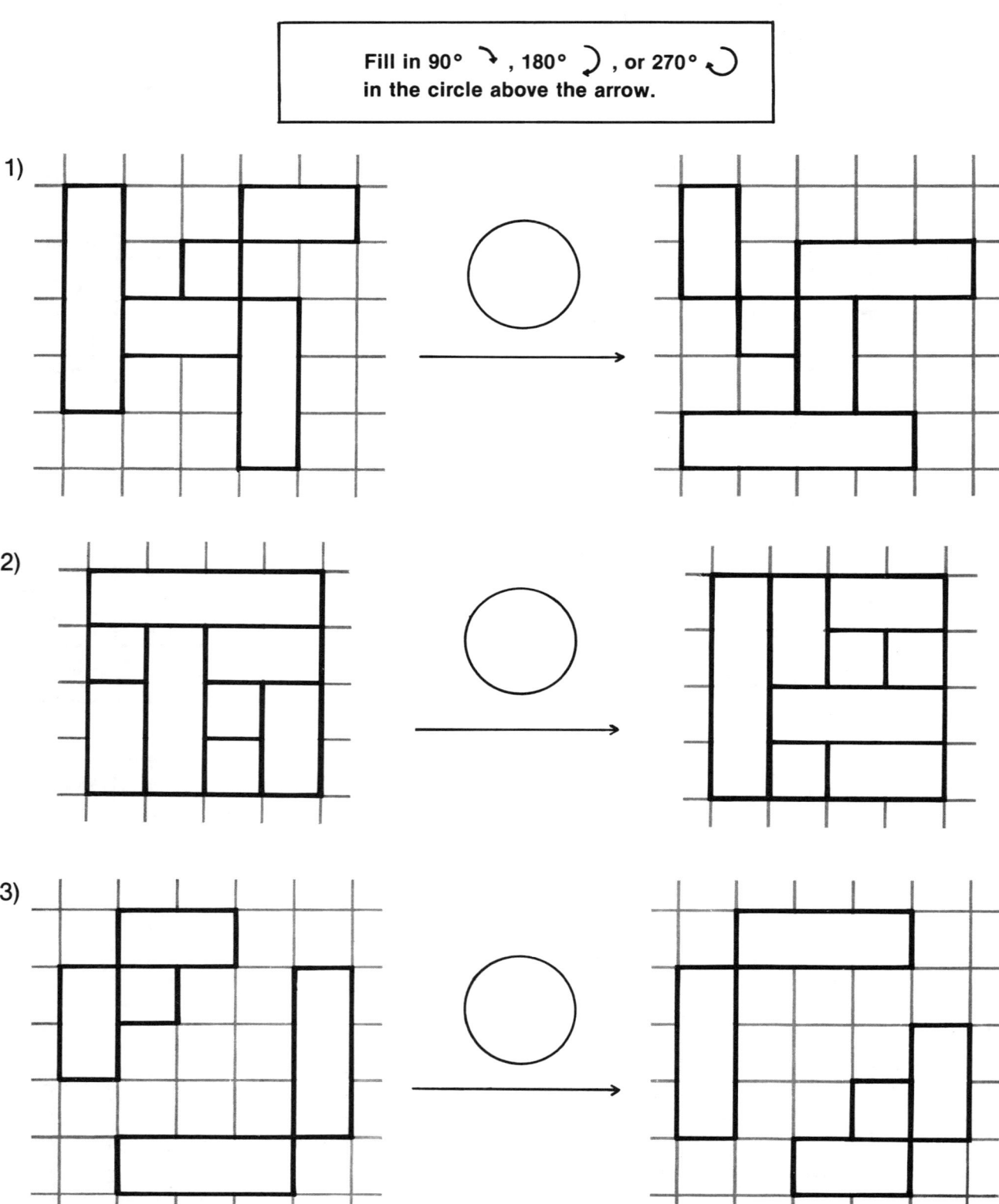

FINDING THE TYPE OF ROTATION

Use rods to make the design shown in the left column. Then name the type of rotation that would produce the resulting design shown in the right column.

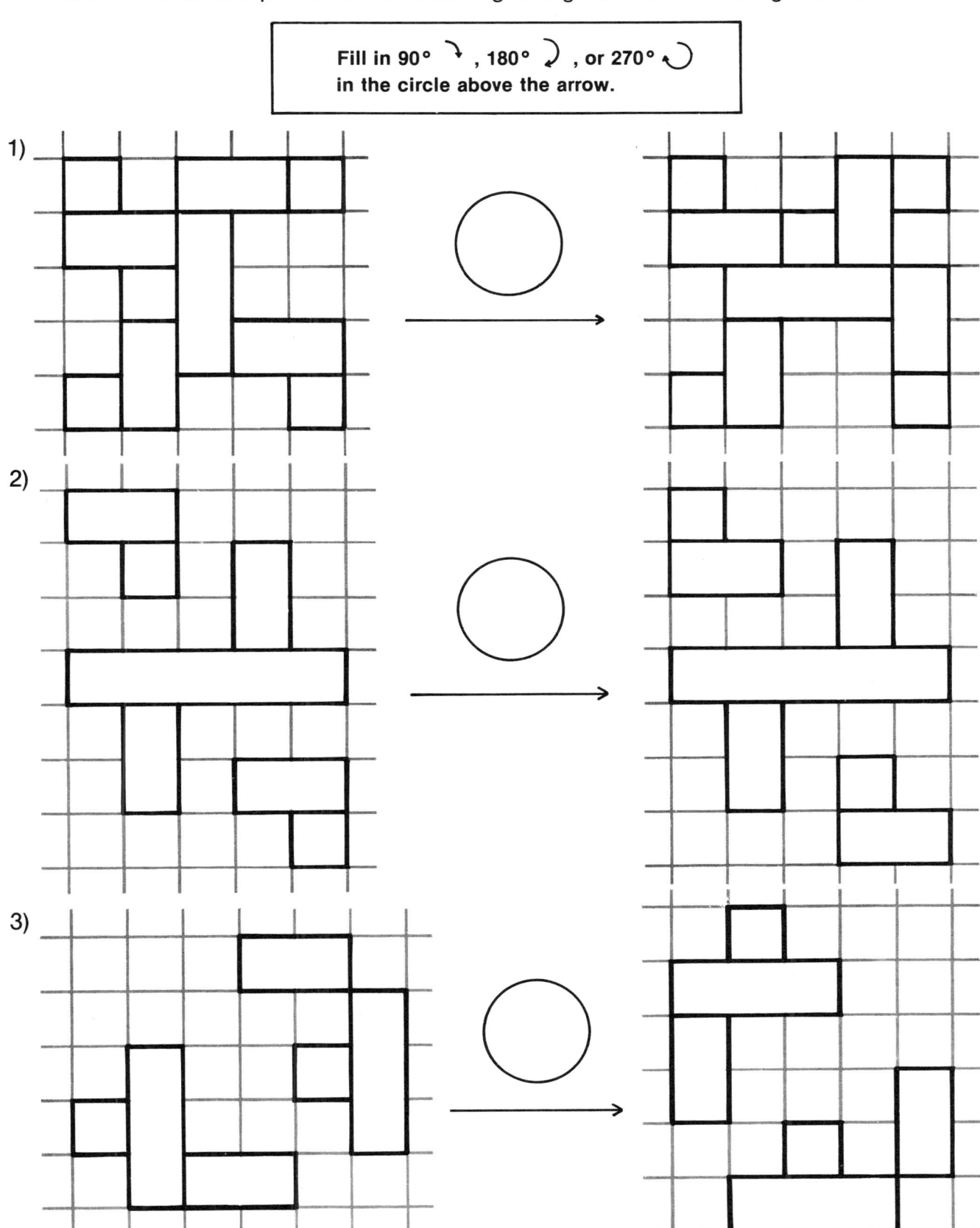

REFLECTING ROD DESIGNS

This page illustrates what a design looks like after it has been reflected in a particular way. For each example, use rods to make the design shown in the left column. Then do the indicated reflection to produce the resulting design in the right column. Compare the original design and the reflected design so that you can visualize each type of reflection. You may wish to use the Master Cut-out #3 from page 60 to help you.

Example of a Vertical Reflection

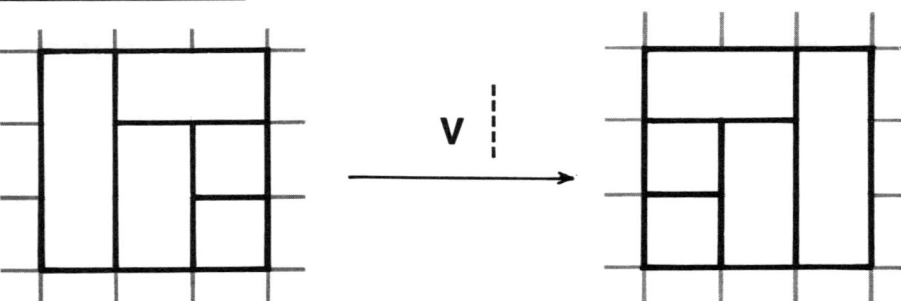

Example of a Horizontal Reflection

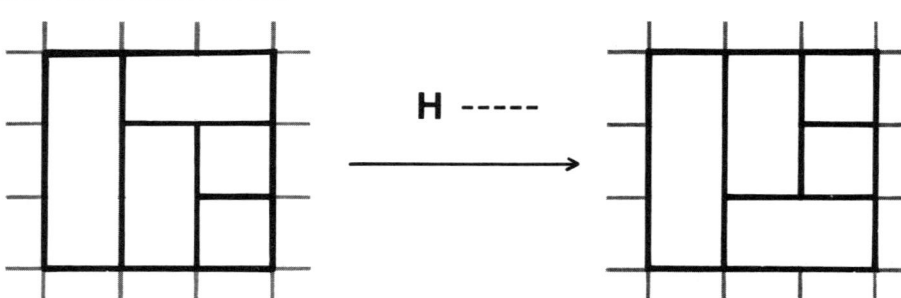

Example of a Left Diagonal Reflection

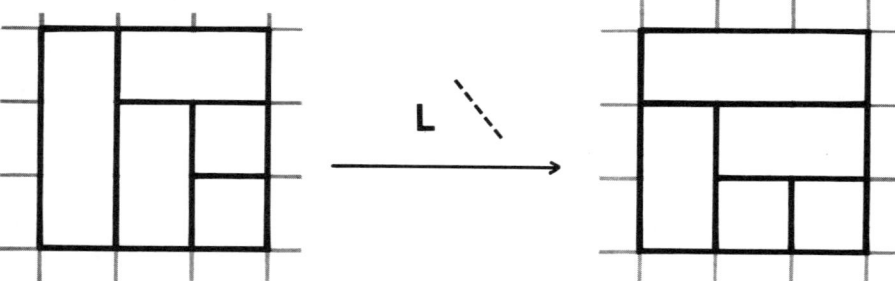

Example of a Right Diagonal Reflection

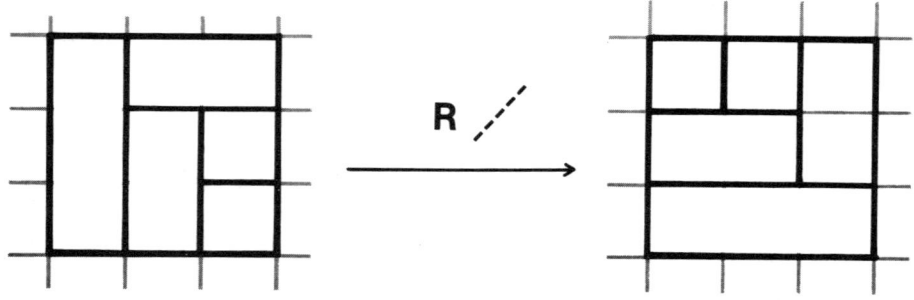

DRAWING REFLECTED ROD DESIGNS

Make the given designs with rods. Reflect each design as indicated. Then draw the design resulting from each reflection. Compare the original design and the reflected design so that you can visualize the effect of each type of reflection. You may wish to use the Master Cut-out #4 from page 60 to help you.

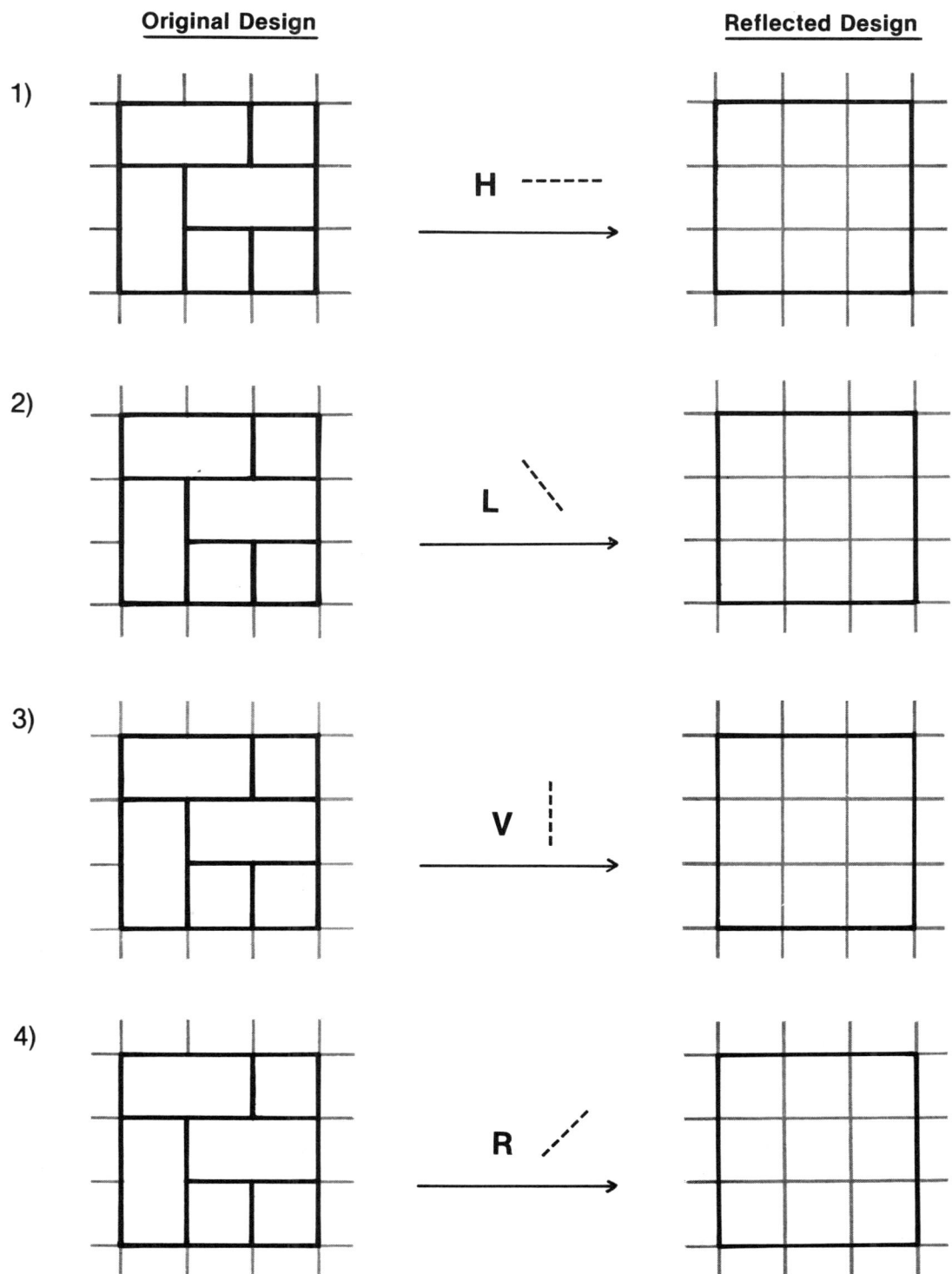

DRAWING REFLECTED ROD DESIGNS

Make the given designs with rods. Reflect each design as indicated. Then draw the design resulting from each reflection. You may wish to use the Master Cut-out #5 to help you.

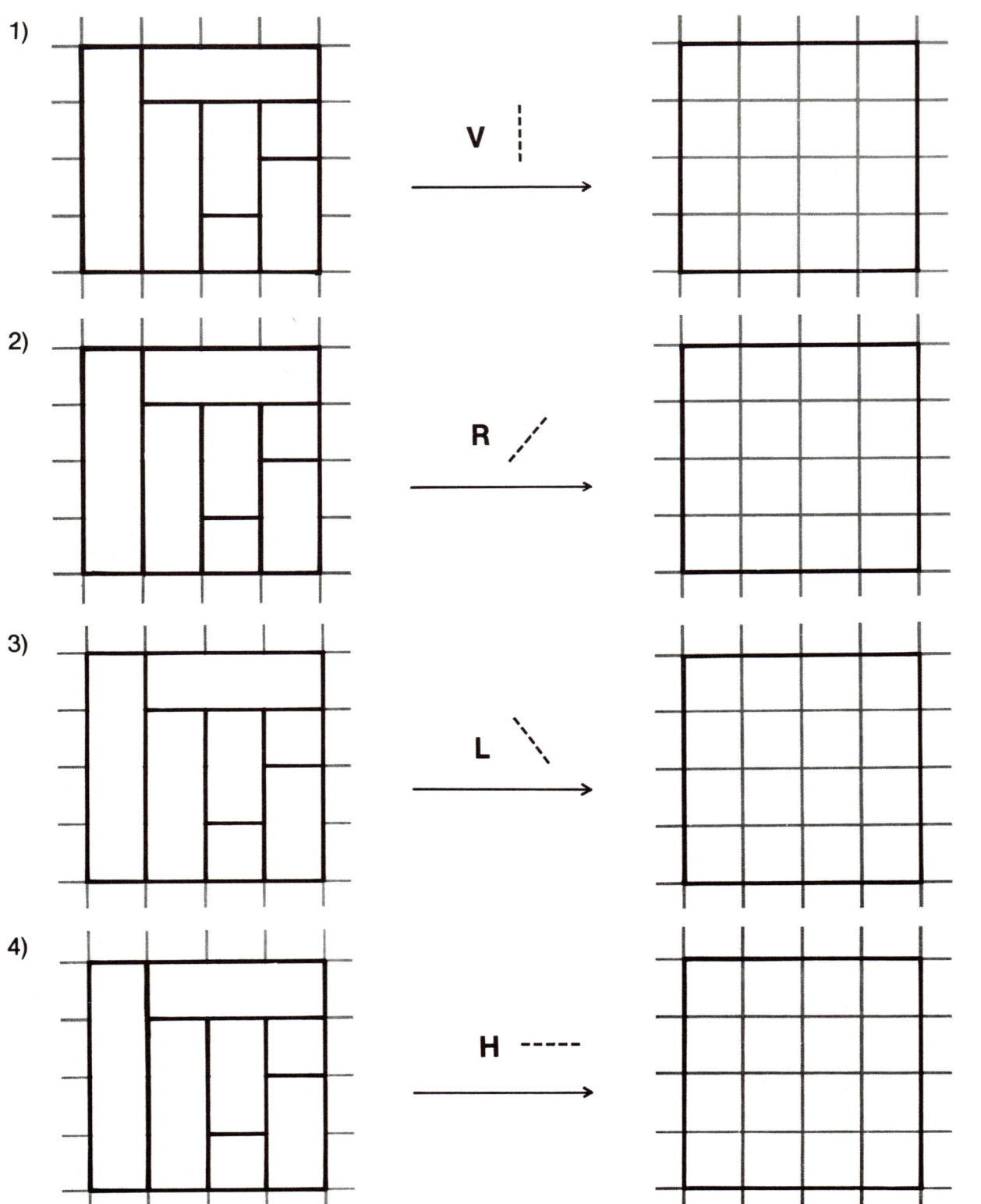

REFLECTING DIFFERENT ROD DESIGNS

Make the given designs with rods. Reflect each design as indicated. Then draw the design resulting from each reflection.

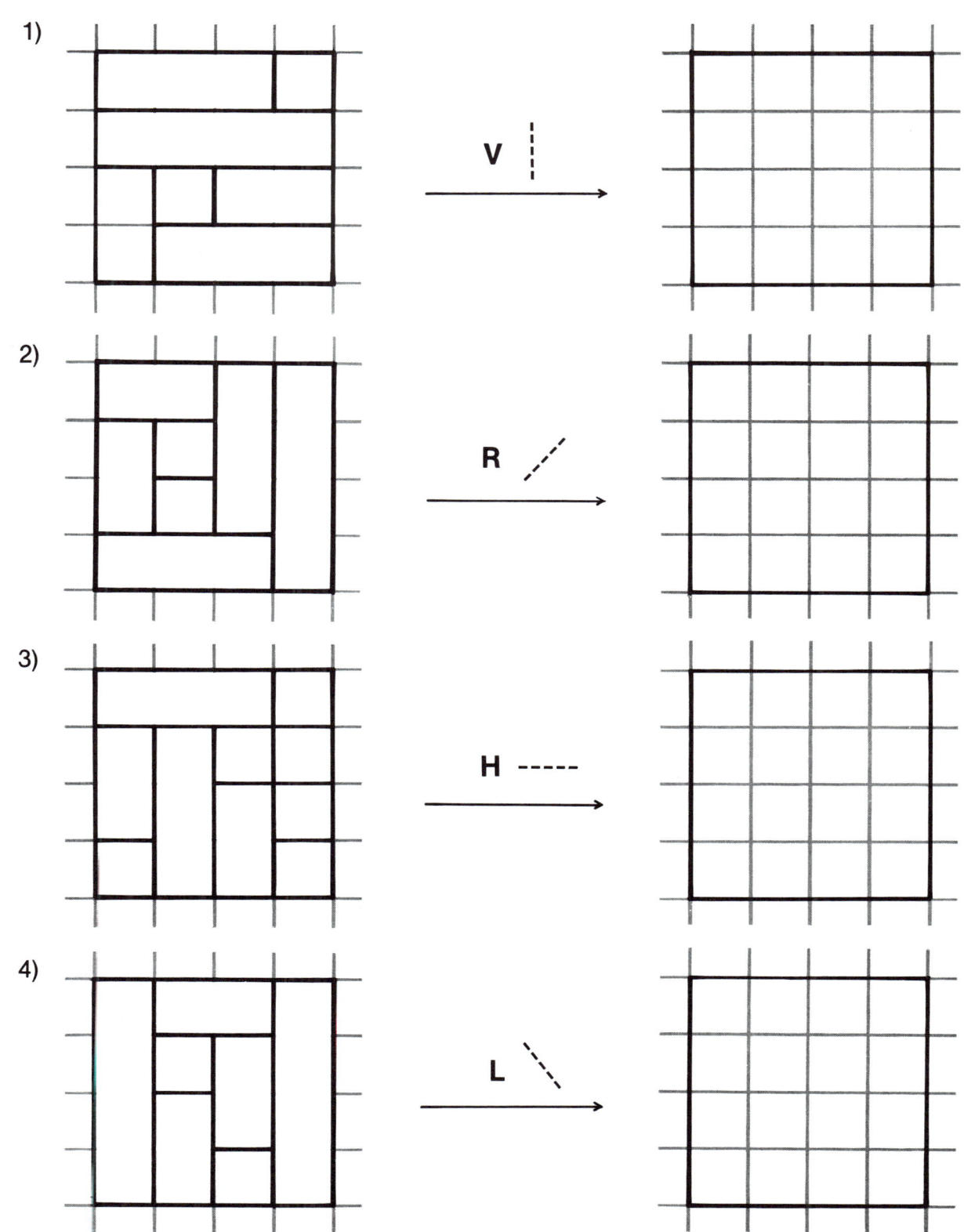

Spatial Problem Solving with Cuisenaire Rods © 1983 Cuisenaire Co. of America, Inc.

FINDING THE TYPE OF REFLECTION

Use rods to make the design shown in the left column. Then name the type of reflection that would produce the resulting design in the right column.

Fill in a V ┊ , H ---- , L ╲ , or R ╱ in the circle above the arrow.

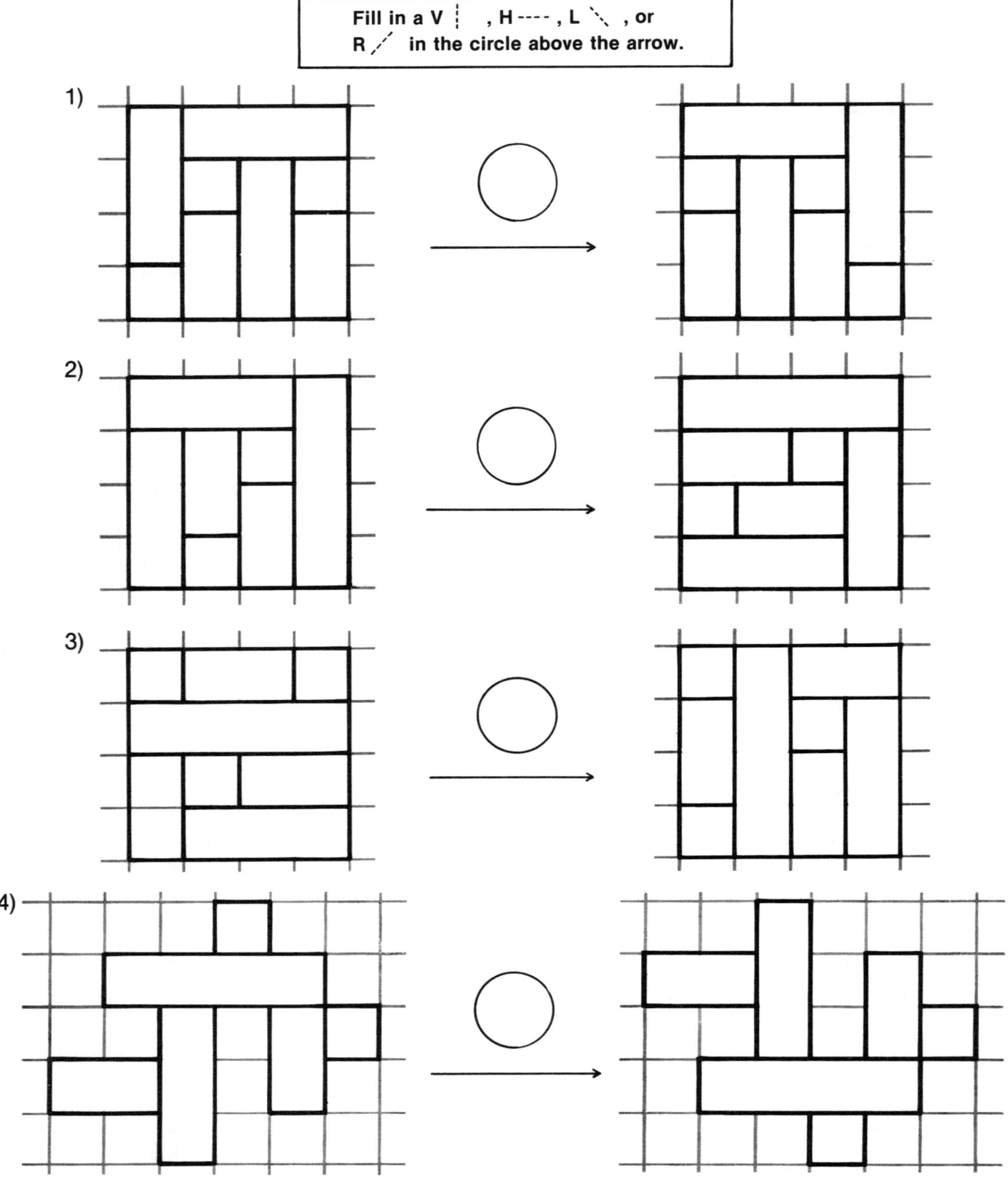

FINDING THE TYPE OF REFLECTION

Use rods to make the design shown in the left column. Then name the type of reflection that would produce the resulting design in the right column.

Fill in a V ¦ , H --- , L \ , or R ╱ in the circle above the arrow.

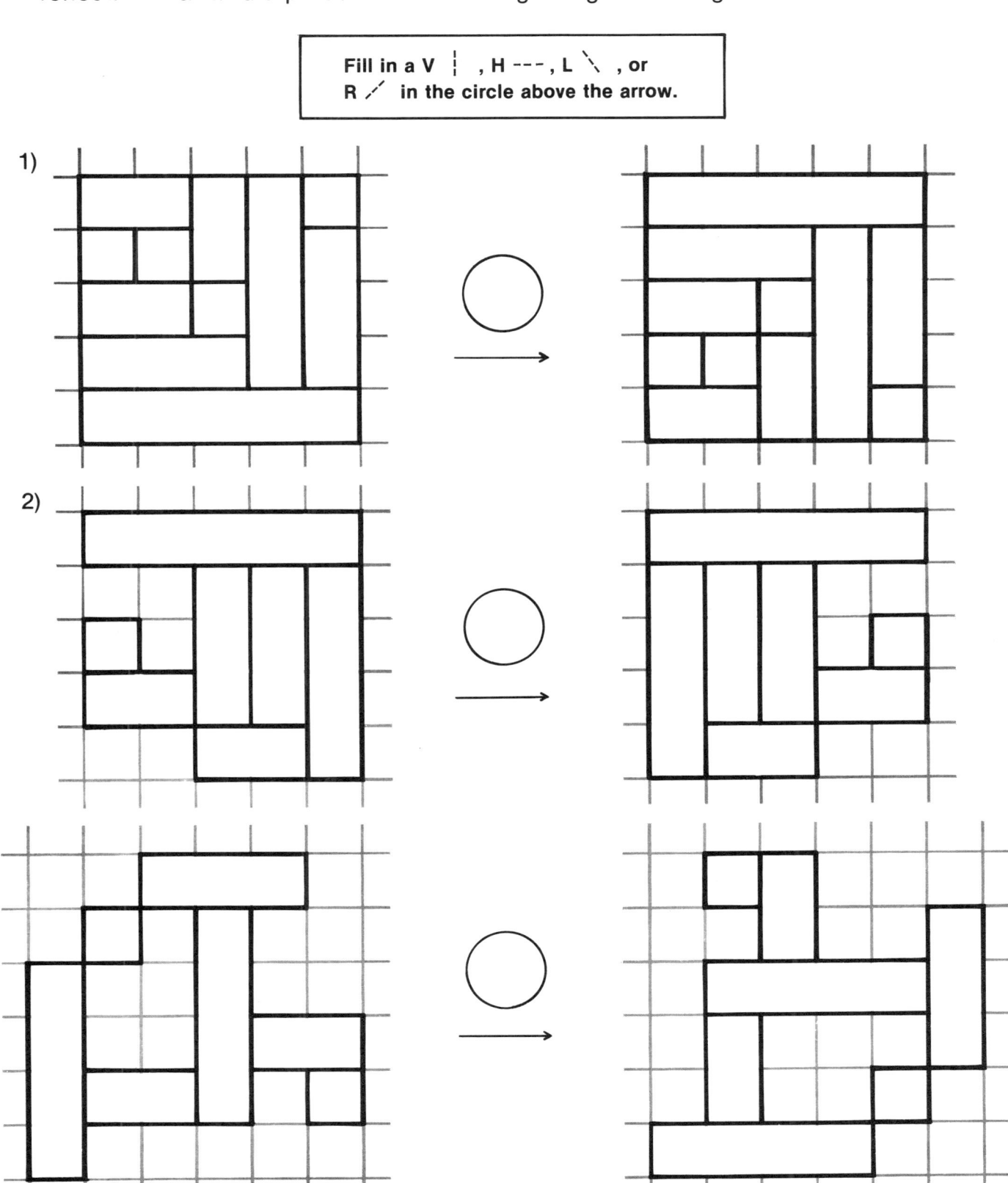

LEARNING HOW TO SOLVE ROD ANALOGY PUZZLES

This rod analogy puzzle should be read as follows:

"The first design is to its resulting design as the second design is to what?"

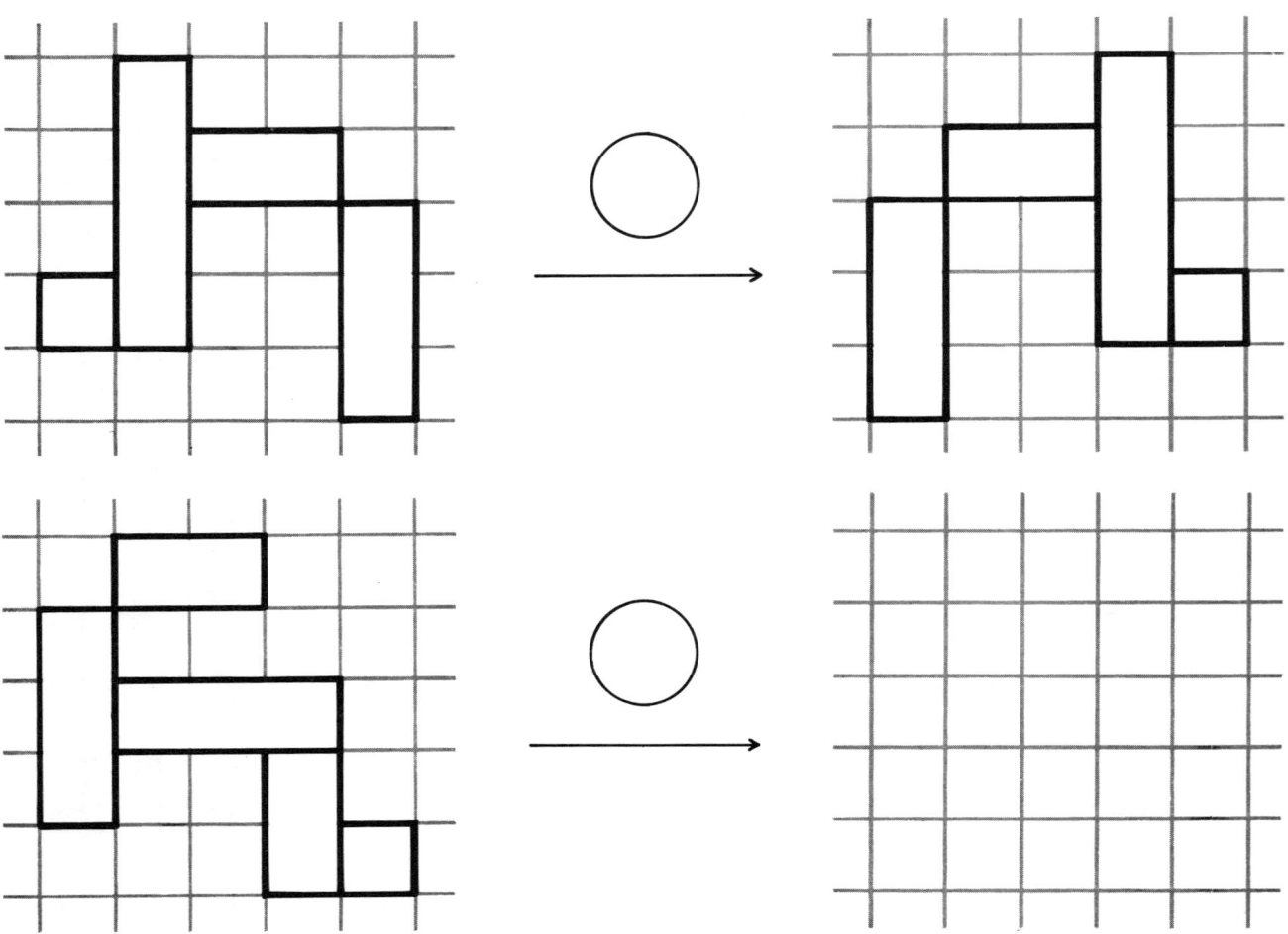

Here are some steps you may wish to use to solve this analogy puzzle:

a) Use rods to make the two given designs on the left.
b) Name the type of rotation or reflection that would produce the given resulting design on the right. (Fill in the code: 90°↱, 180°↻, 270°↺, V⋮, H⋯, L↘ or R↗ in the circle over the first arrow.)
c) Now do that same rotation or reflection to the second design on the left. Fill in the second circle with the same code.
d) Find and draw the resulting design in the space provided.

Did you find that you needed to do a vertical reflection, V⋮?

Page 27   Spatial Problem Solving with Cuisenaire Rods © 1983 Cuisenaire Co. of America, Inc.

SOLVING ROTATION AND REFLECTION ANALOGY PUZZLES

Solve each of these three analogy puzzles using your rods. Fill in the circles with the code: 90°↷, 180°↻, 270°↺, V |, H----, L↘, or R↗. Find and draw the missing resulting design in the space provided.

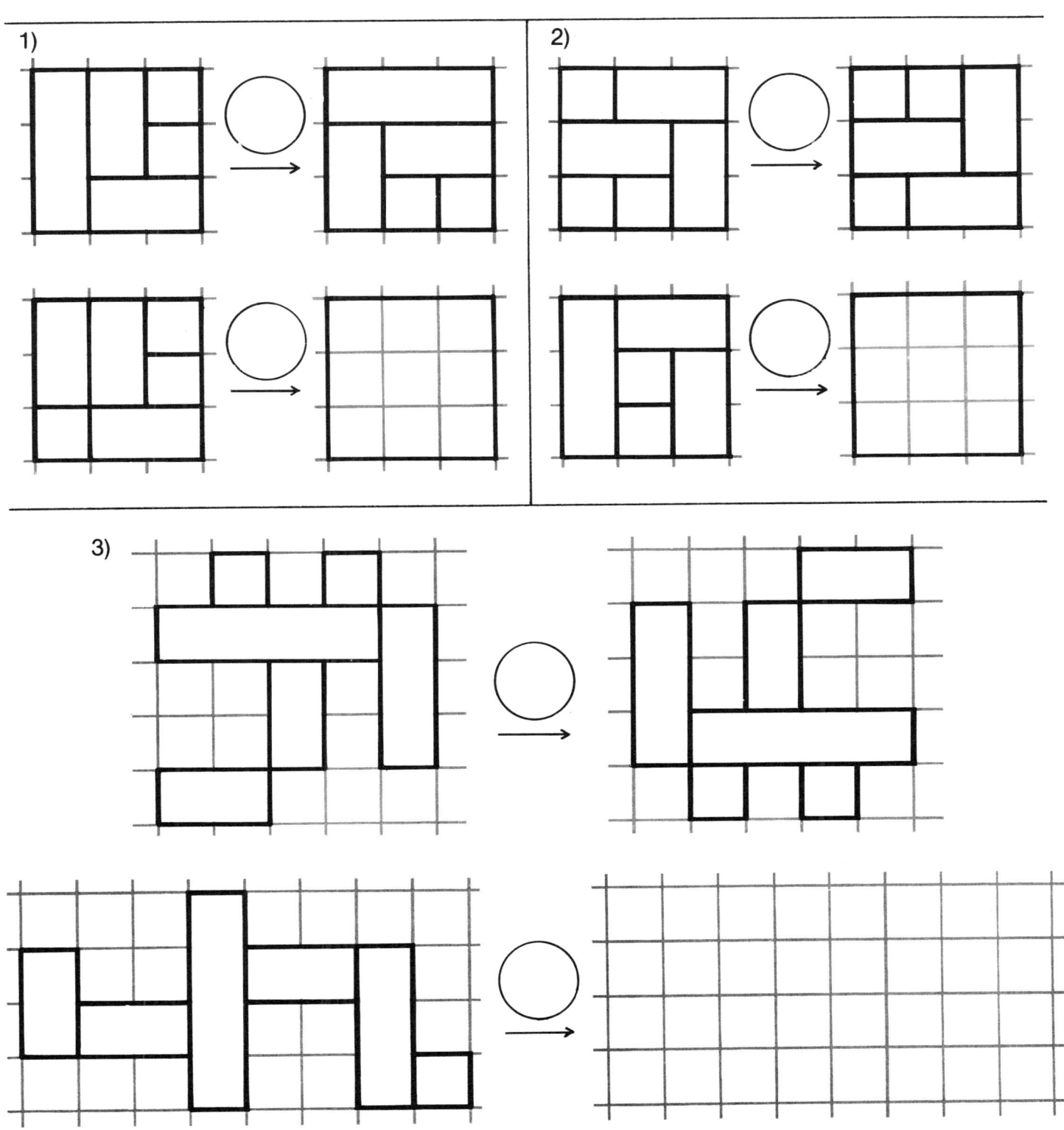

Spatial Problem Solving with Cuisenaire Rods © 1983 Cuisenaire Co. of America, Inc.

SOLVING ROTATION AND REFLECTION ANALOGY PUZZLES

Solve each of these three analogy puzzles using your rods. Fill in the circles with the code: 90°↷, 180°↺, 270°↻, V|, H---, L\\, or R/. Find and draw the missing resulting design in the space provided.

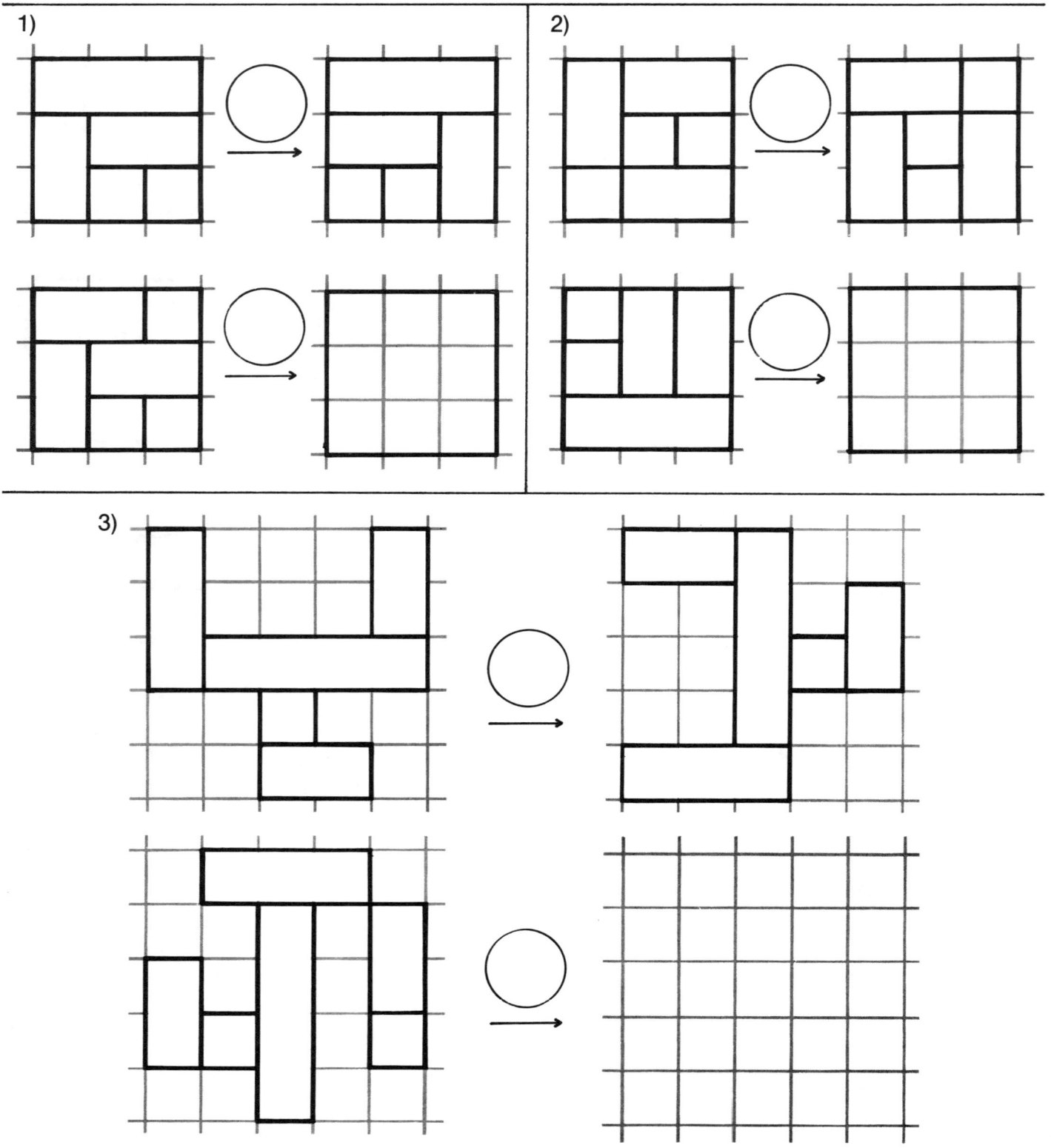

Page 29 Spatial Problem Solving with Cuisenaire Rods © 1983 Cuisenaire Co. of America, Inc.

SOLVING ROTATION AND REFLECTION ANALOGY PUZZLES

Solve each of these three analogy puzzles using your rods. Fill in the circles with the code: 90° ↘, 180° ↩, 270° ↺, V ¦ , H ---, L ↘, or R ↗. Find and draw the missing resulting design in the space provided.

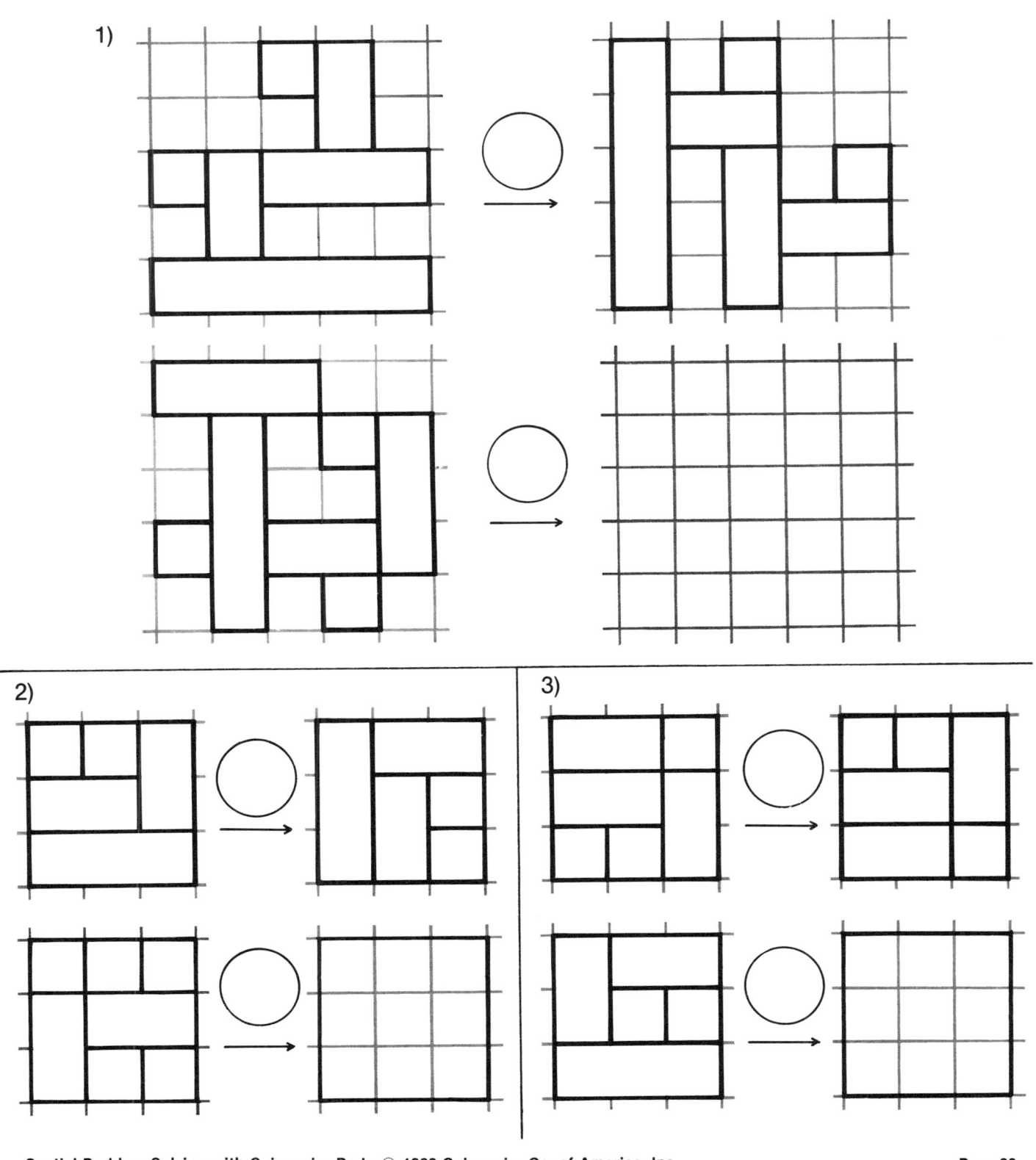

Spatial Problem Solving with Cuisenaire Rods © 1983 Cuisenaire Co. of America, Inc. Page 30

SOLVING ROTATION AND REFLECTION ANALOGY PUZZLES

Solve each of these three analogy puzzles using your rods. Fill in the circles with the code: 90°↷, 180°↻, 270°↶, V |, H ---, L ↘, or R ↗. Find and draw the missing resulting design in the space provided.

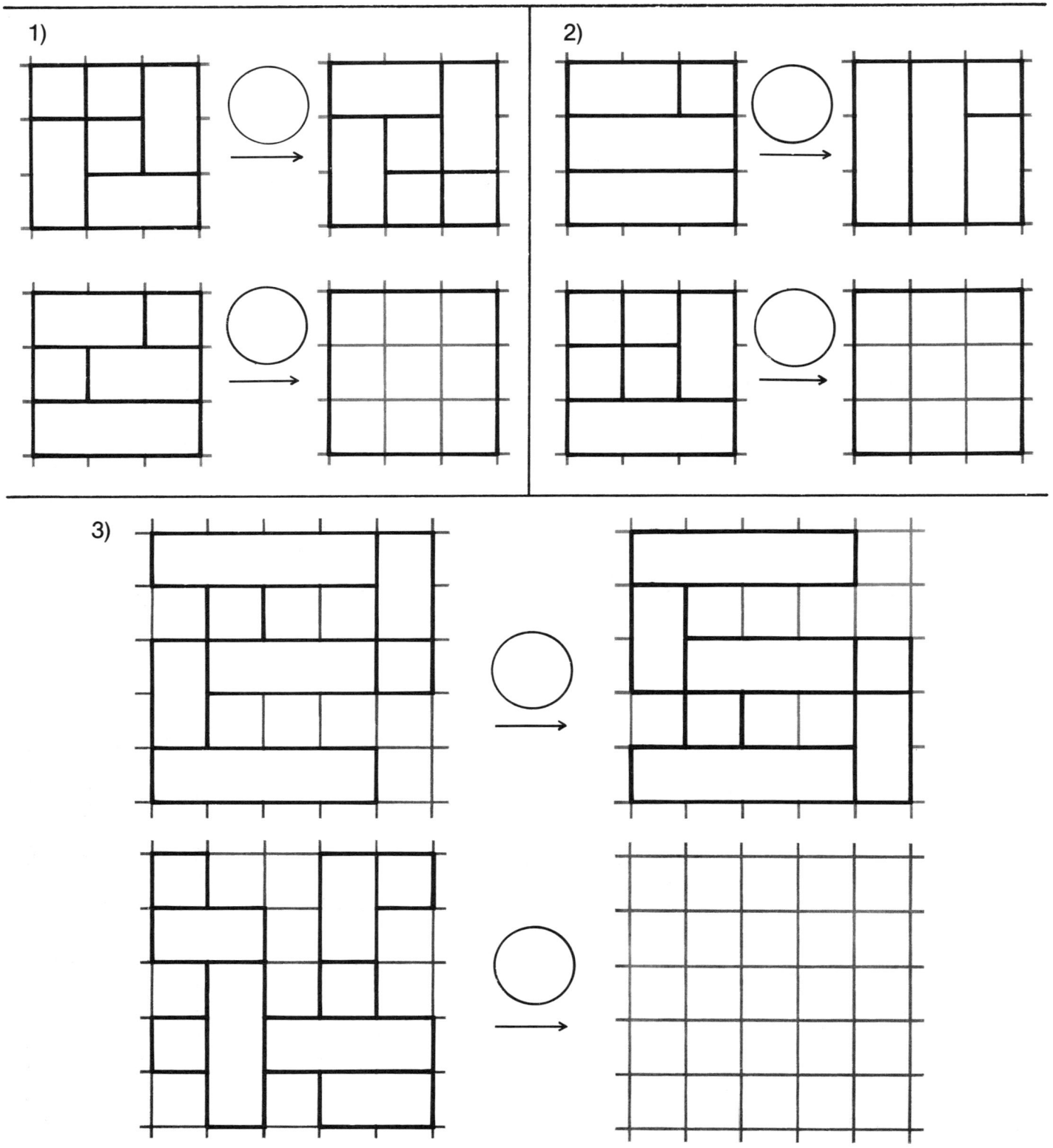

Page 31   Spatial Problem Solving with Cuisenaire Rods © 1983 Cuisenaire Co. of America, Inc.

SOLVING ROTATION AND REFLECTION ANALOGY PUZZLES

Solve each of these three analogy puzzles using your rods. Fill in the circles with the code: 90°↘, 180°⟩, 270°↺, V|, H---, L⟍, or R⟋. Find and draw the missing resulting design in the space provided

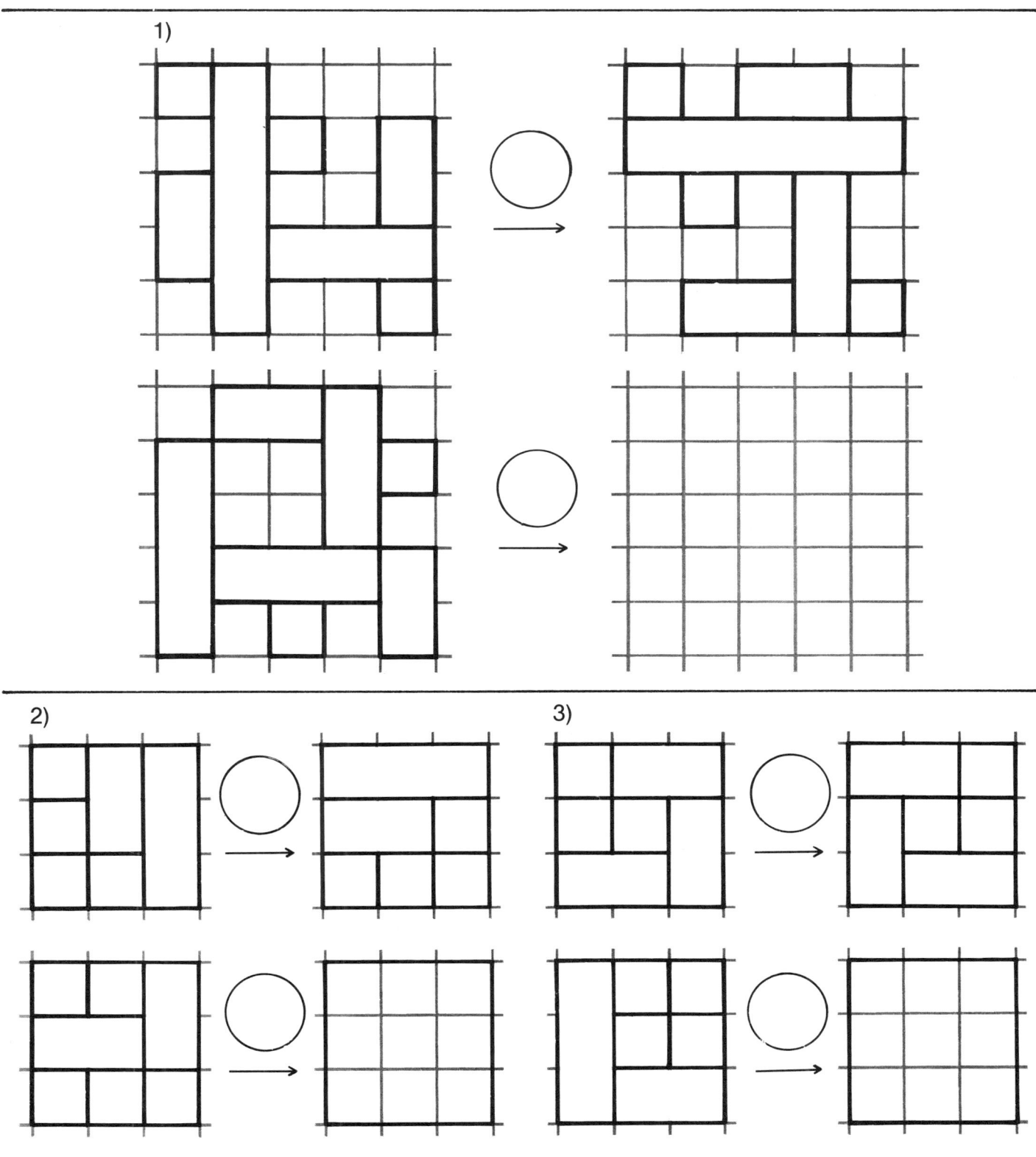

SOLVING ROTATION AND REFLECTION ANALOGY PUZZLES

Solve each of these analogy puzzles using your rods. Fill in the circles with the code: 90°↘, 180°⤵, 270°⤴, V |, H ---, L ╲, or R ╱. Find and draw the missing resulting design in the space provided.

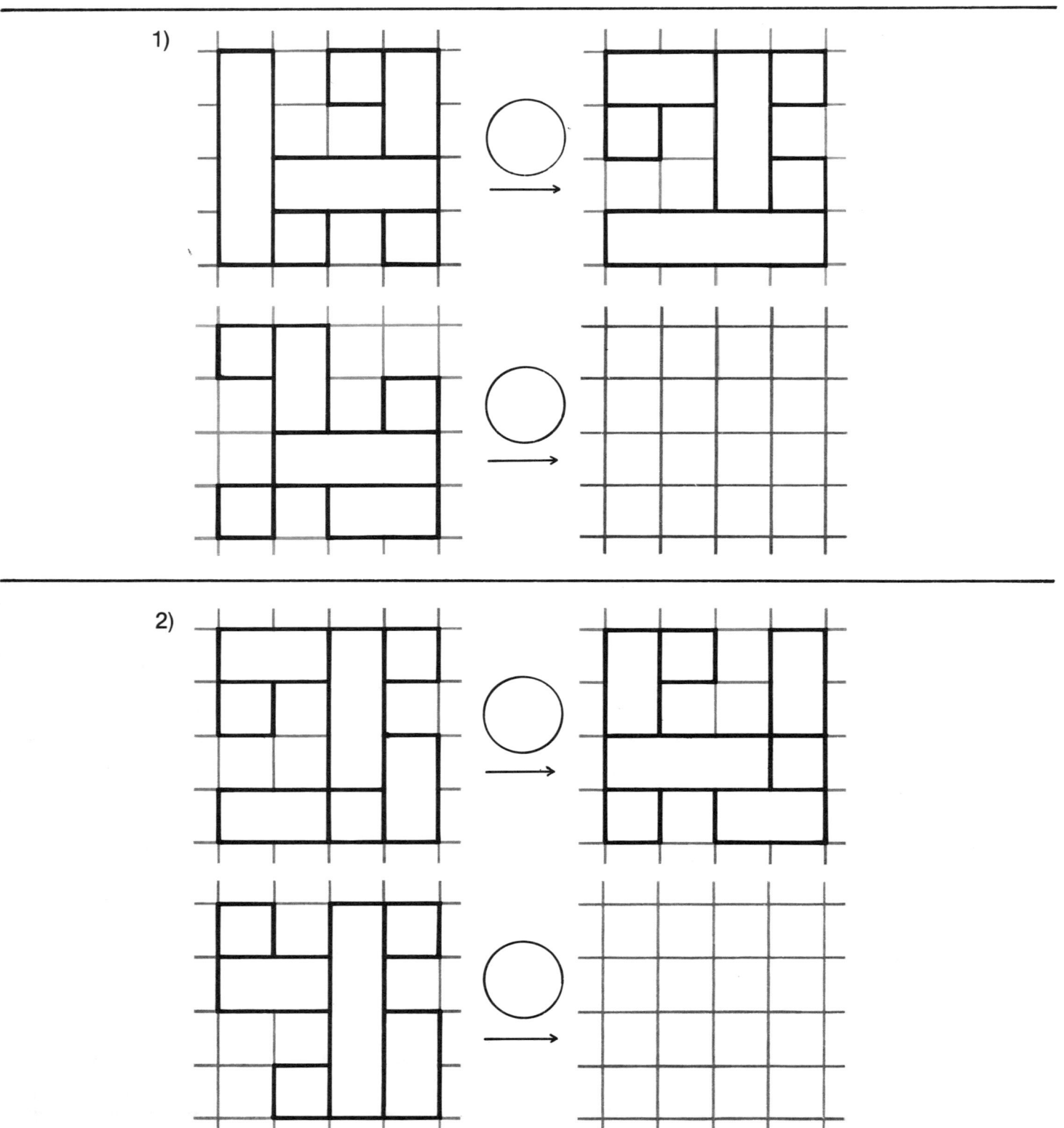

Page 33   Spatial Problem Solving with Cuisenaire Rods © 1983 Cuisenaire Co. of America, Inc.

SOLVING ROTATION AND REFLECTION ANALOGY PUZZLES

Solve each of these analogy puzzles using your rods. Fill in the circles with the code: 90°↘, 180°↻, 270°↺, V|, H---; L↘, or R↗. Find and draw the missing resulting design in the space provided.

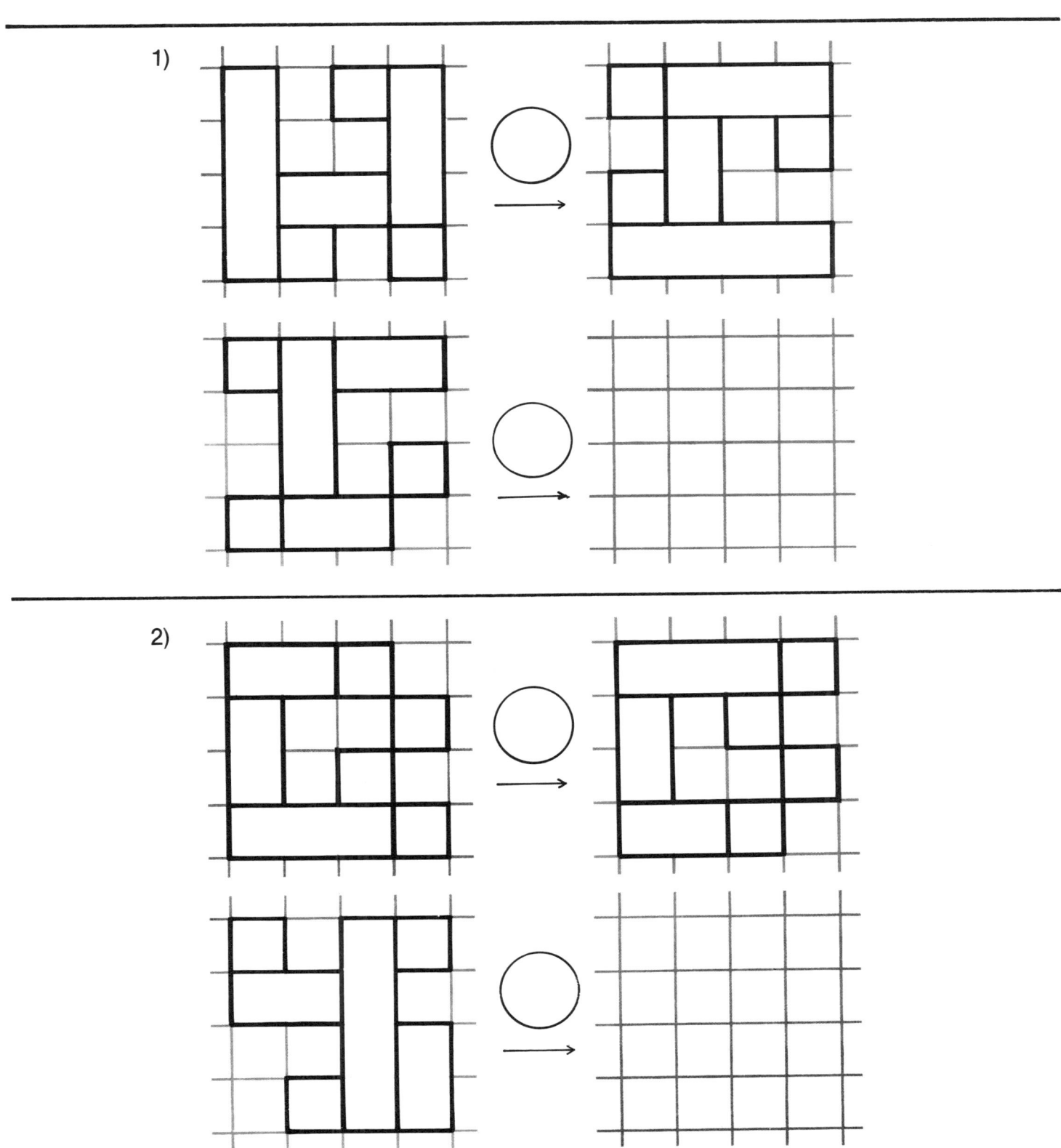

Spatial Problem Solving with Cuisenaire Rods © 1983 Cuisenaire Co. of America, Inc.

REVERSING ROD ANALOGY PUZZLES

Use your rods to find the missing designs to make these correct analogies. Fill in the circles with the code: 90°↘, 180°↩, 270°↪, V ┊, H ┄, L↘, or R↗. Draw the missing design in the space provided.

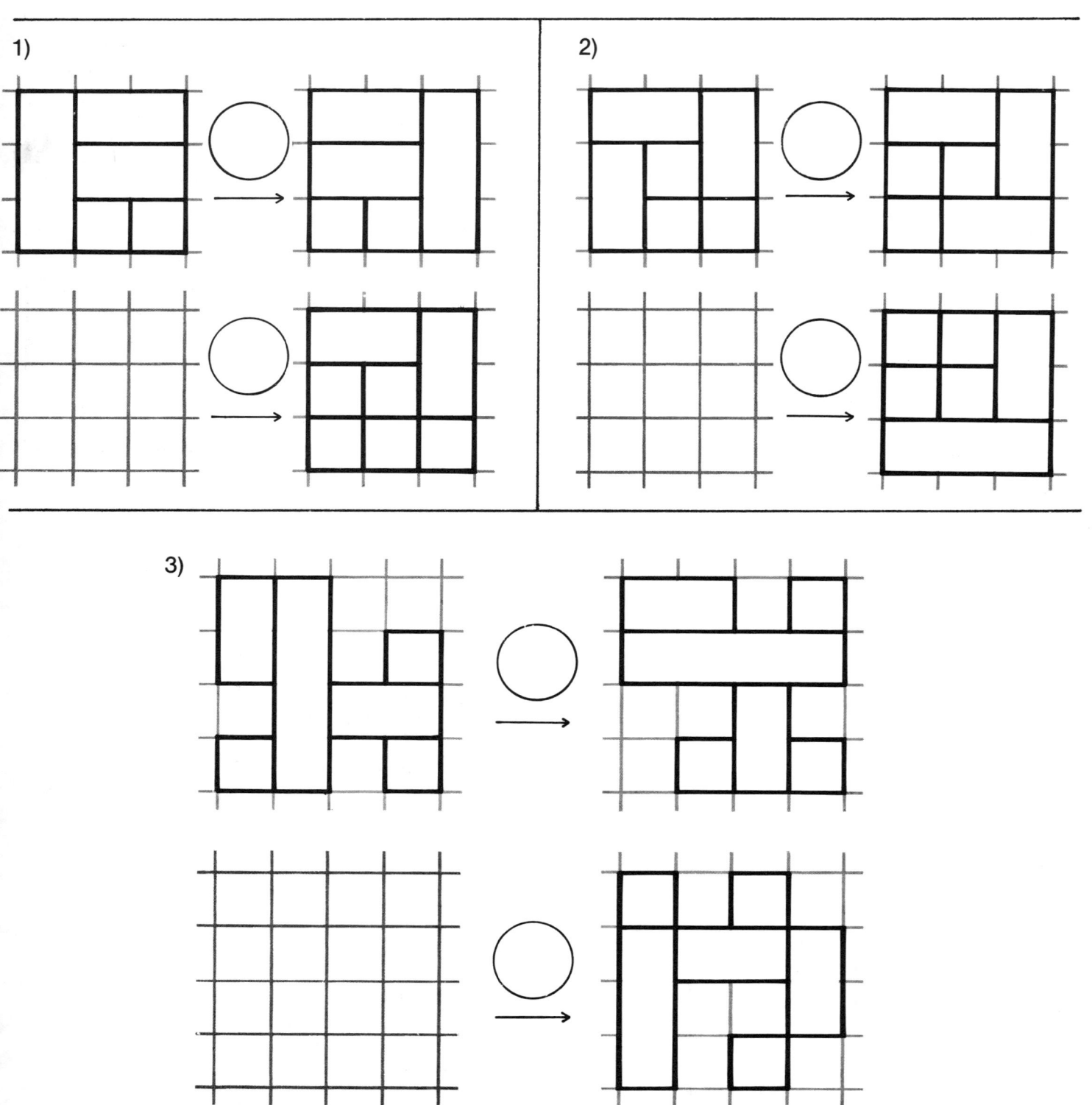

REVERSING ROD ANALOGY PUZZLES

Use your rods to find the missing designs to make these correct analogies. Fill in the circles with the code: 90°↘, 180°↷, 270°↺, V ┊, H ┄, L ╲, or R ╱. Draw the missing design in the space provided.

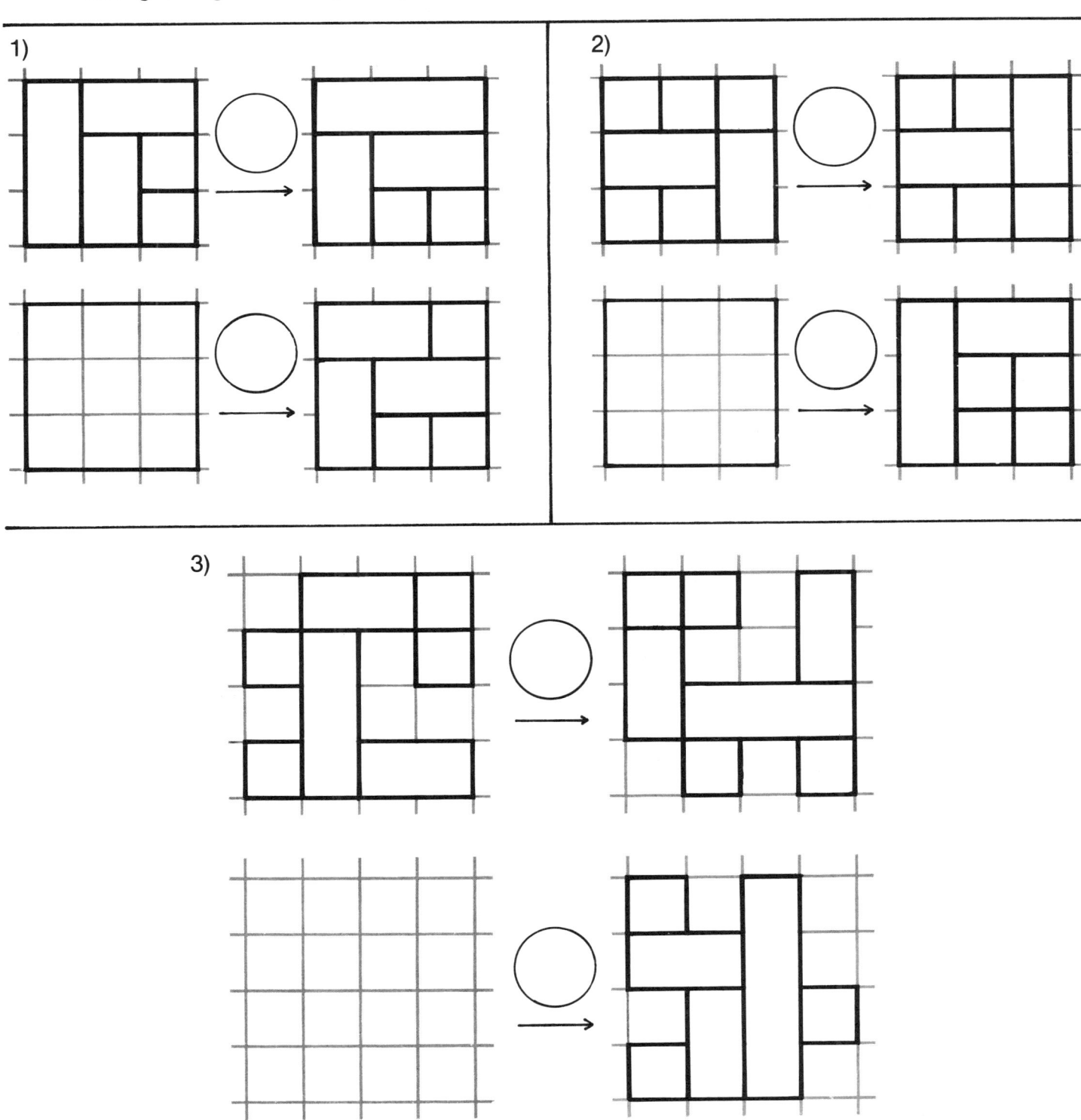

Spatial Problem Solving with Cuisenaire Rods © 1983 Cuisenaire Co. of America, Inc.

SOLVING A CHAIN OF ROTATIONS AND REFLECTIONS

Use rods to make the design at the top of the page. Follow the arrows and perform the chain of rotations and reflections in succession. Draw (in pencil) the resulting design at each stage until you get all the way down the left side of the page and back up the right side of the page. You should end with your original design. If you didn't, try again! You may wish to use the Master Cut-out #6 from page 60 to help you.

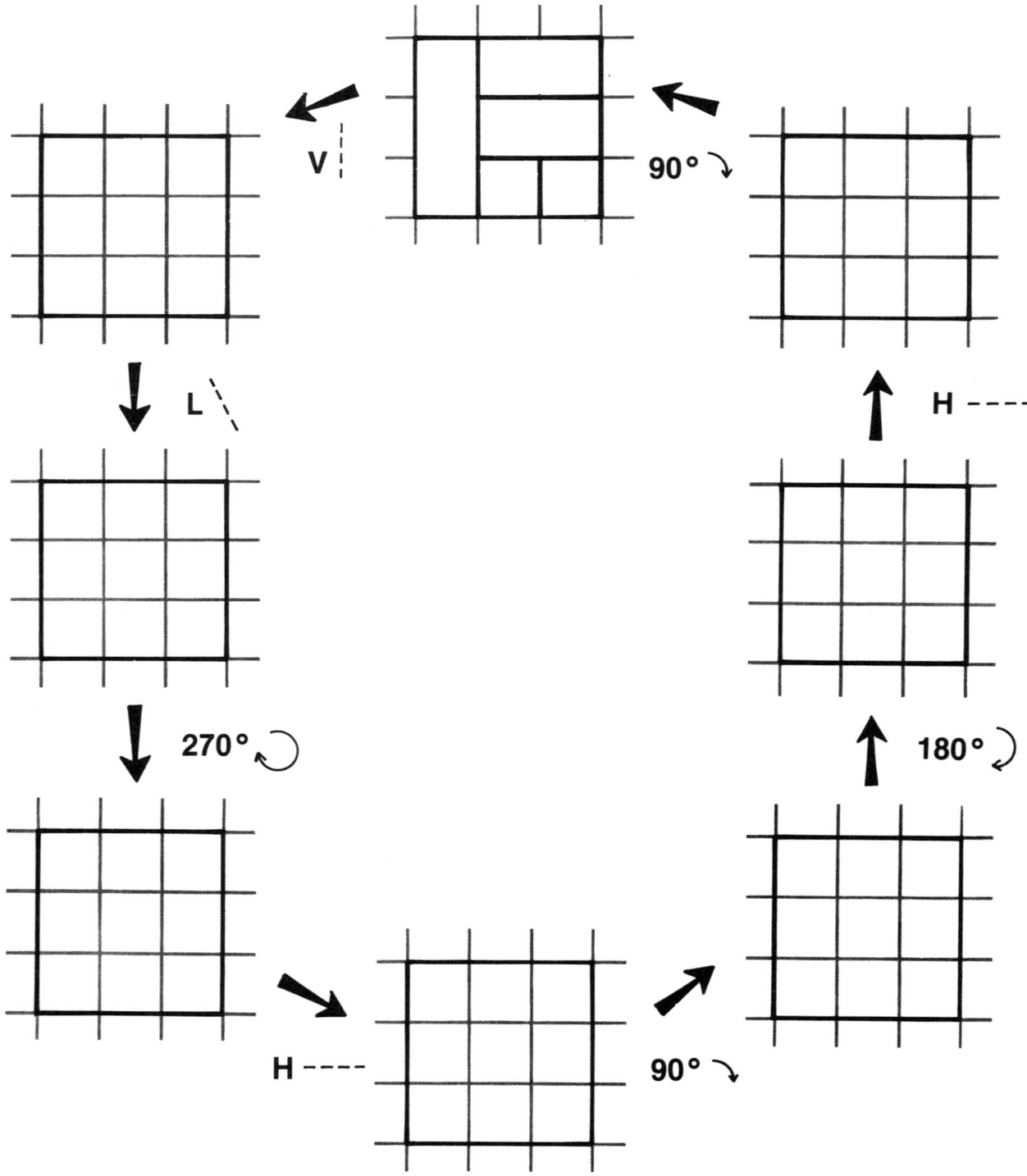

SOLVING A CHAIN OF ROTATIONS AND REFLECTIONS

Use rods to make the design at the top of the page. Follow the arrows and perform the chain of rotations and reflections in succession. Your should end with your original design. If you didn't, try again.

Draw (in pencil) the resulting design at each stage.

You may wish to use the Master Cut-out #7 from page 60 to help you.

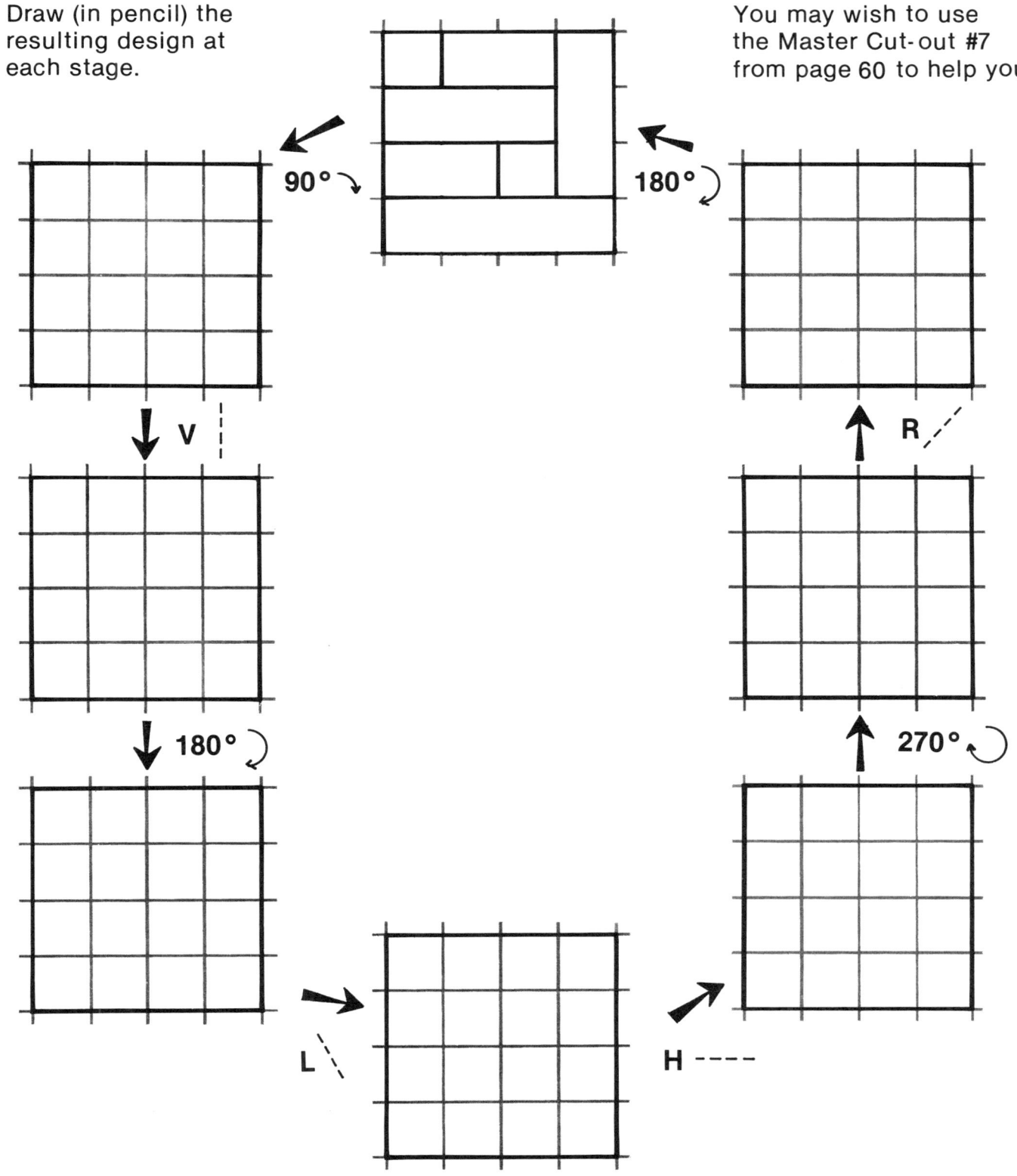

Spatial Problem Solving with Cuisenaire Rods © 1983 Cuisenaire Co. of America, Inc.

SOLVING A CHAIN OF ROTATIONS AND REFLECTIONS

Use rods to make the design at the top of the page. Follow the arrows and perform the chain of rotations and reflections in succession. You should end with your original design. If you didn't, try again.

Draw (in pencil) the resulting design at each stage.

You may wish to use the Master Cut-out #8 from page 60 to help you.

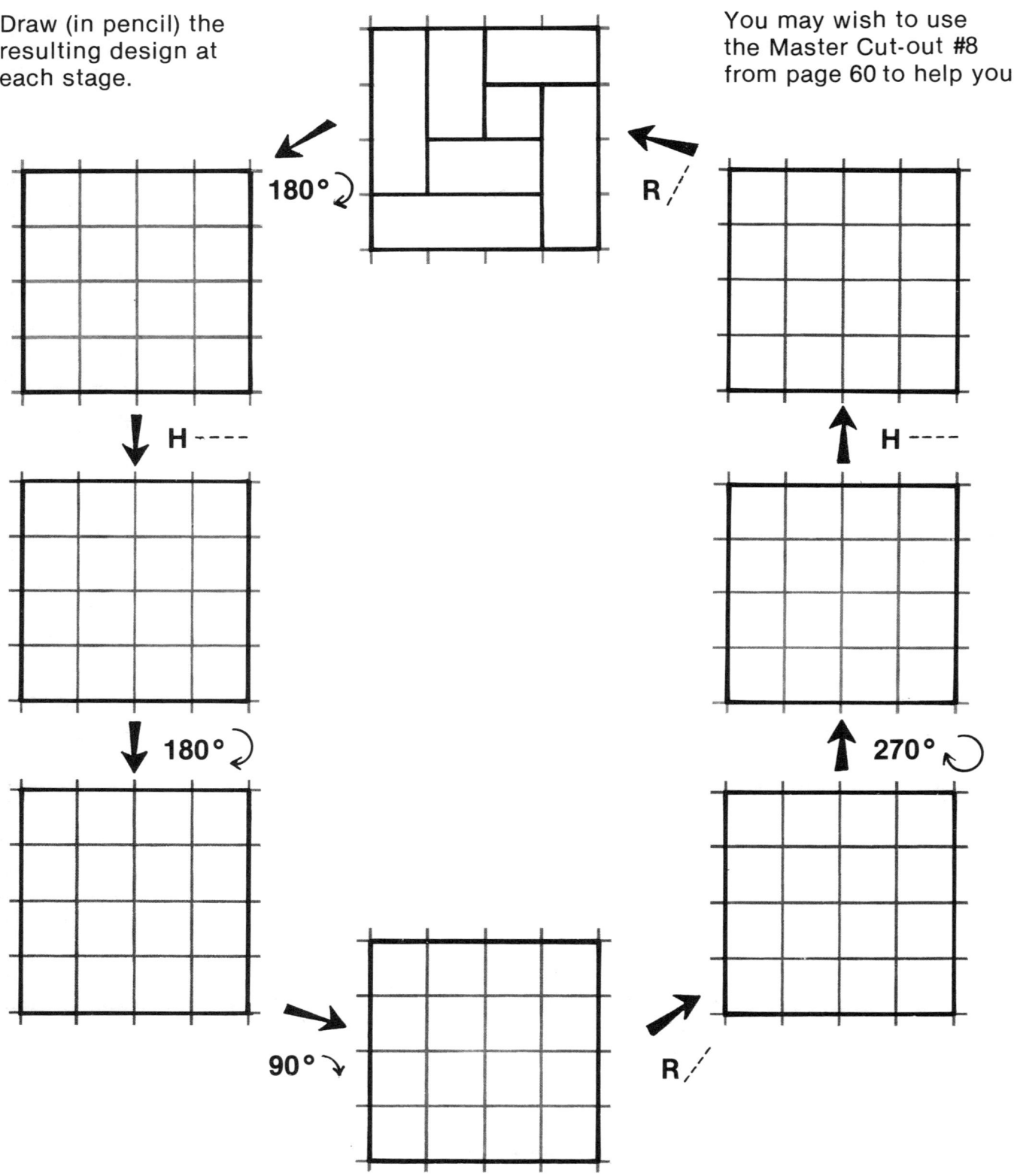

Page 39   SPATIAL PROBLEM SOLVING WITH Cuisenaire Rods © 1983 Cuisenaire Co. of America, Inc.

SOLVING A CHAIN OF ROTATIONS AND REFLECTIONS

Use rods to make the design at the top of the page. Follow the arrows and perform the chain of rotations and reflections in succession. You should end with your original design. If you didn't, try again.

Draw (in pencil) the resulting design at each stage.

You may wish to use the Master Cut-out #9 from page 60 to help you.

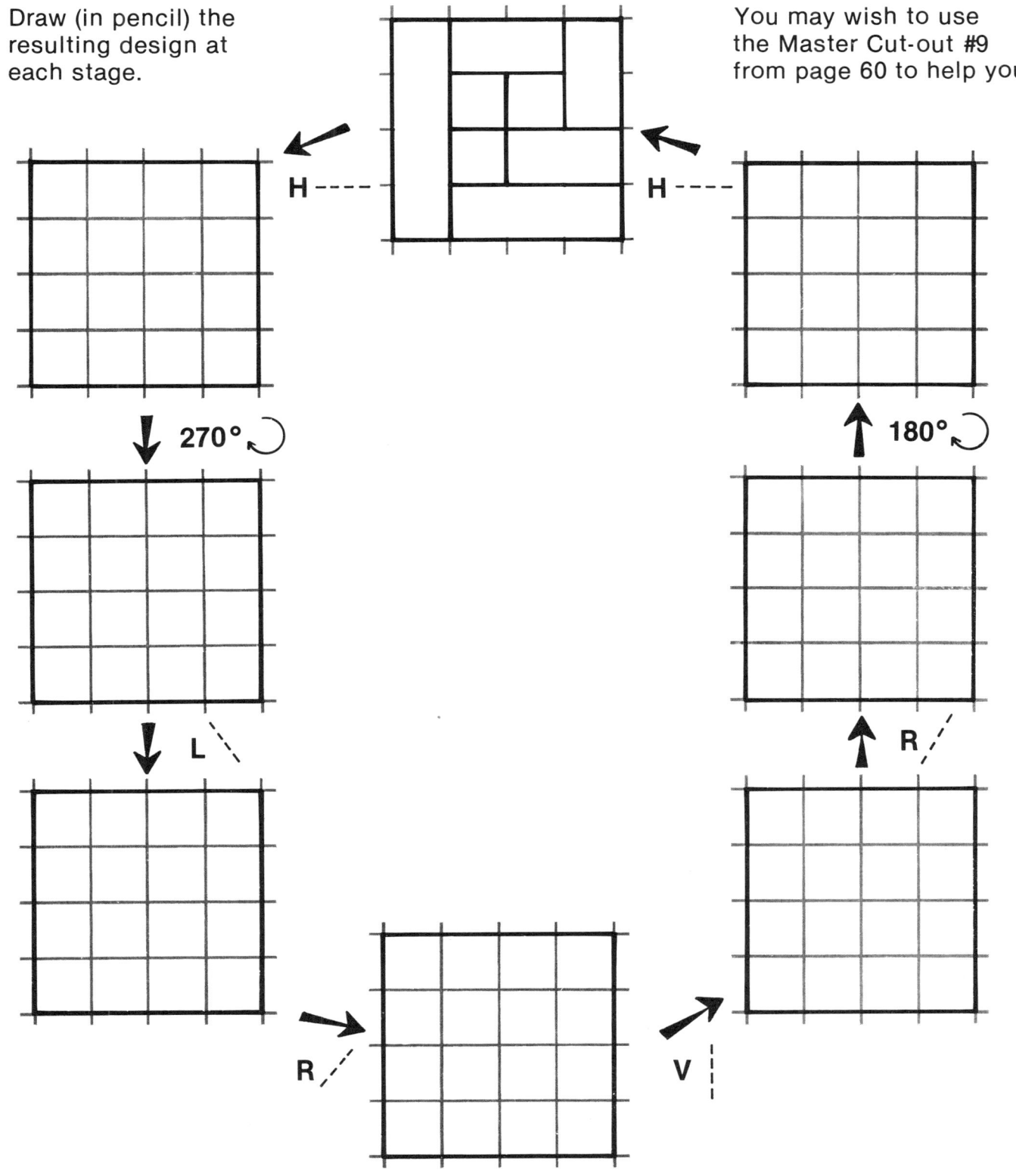

Spatial Problem Solving with Cuisenaire Rods © 1983 Cuisenaire Co. of America, Inc.

Page 40

SEEING THREE-DIMENSIONAL ROD DESIGNS IN TWO DIMENSIONS

1) Build this three-dimensional rod design using 1 yellow, 1 light green, 1 red, and 1 white rod.

Pictured here are the two-dimensional top, front, and side views of the rod design. Observe carefully how they are obtained.

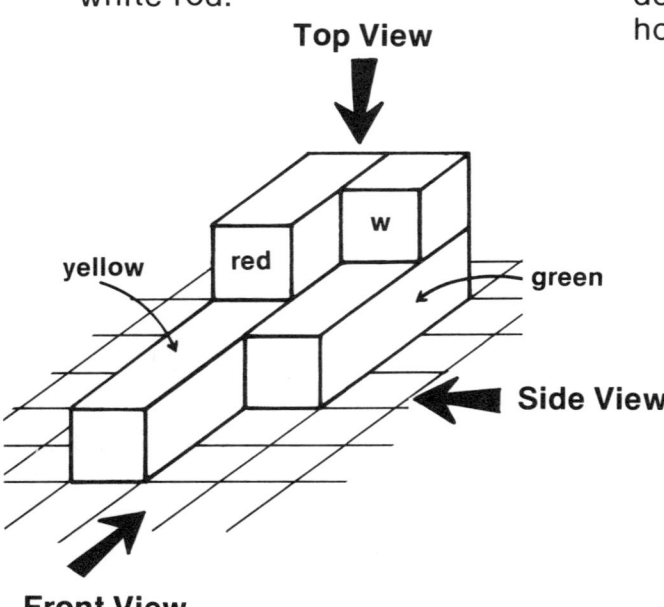

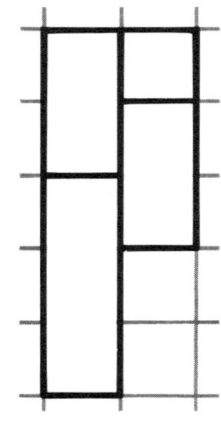

2) Now build this three-dimensional rod design using 1 yellow, 1 purple, 1 light green and 1 white rod.

Observe its top, front, and side views.

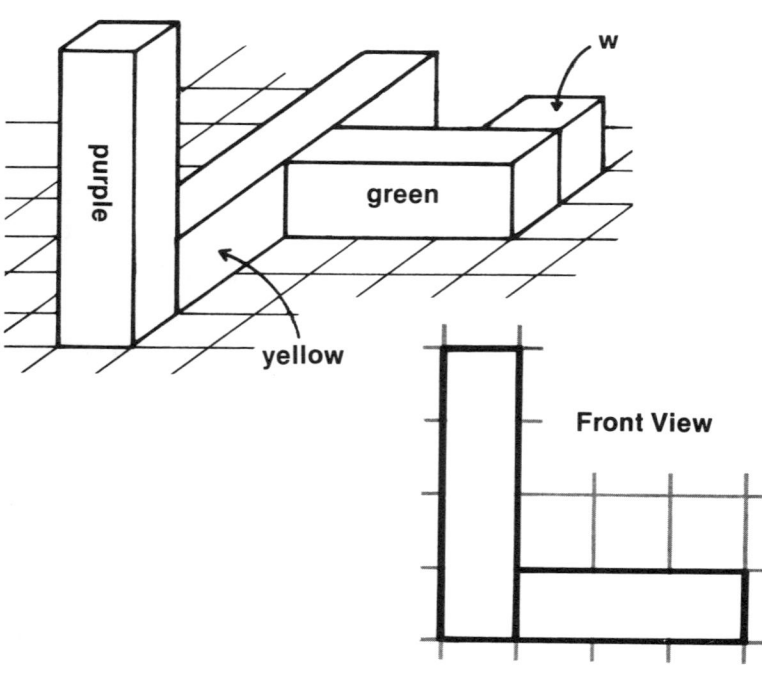

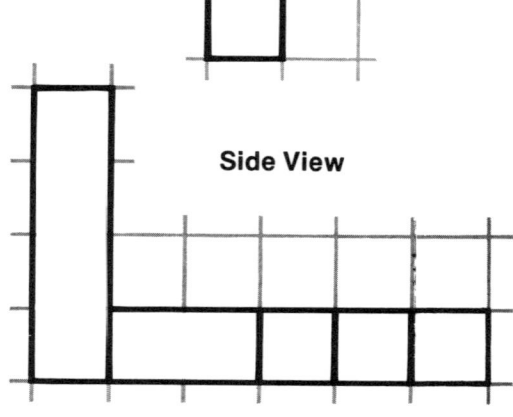

Spatial Problem Solving with Cuisenaire Rods © 1983 Cuisenaire Co. of America, Inc.

DRAWING THREE-DIMENSIONAL ROD DESIGNS IN TWO DIMENSIONS

1) Build this three-dimensional rod design using 2 light green, 2 red, and 1 white rod.

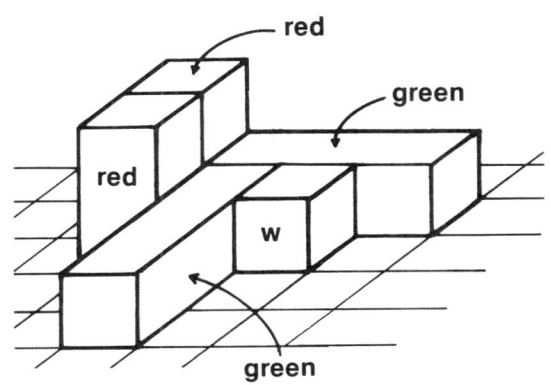

Draw the top view. The front and side views are given.

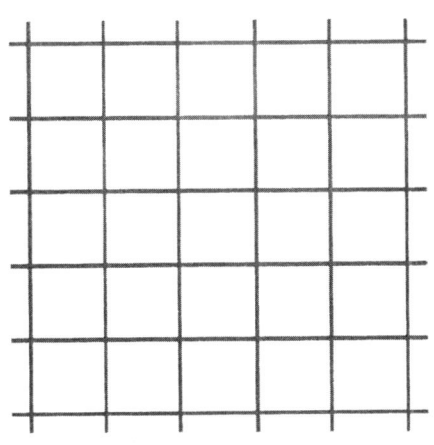

Top View

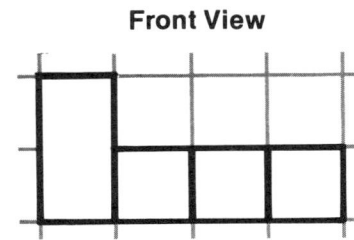

Front View

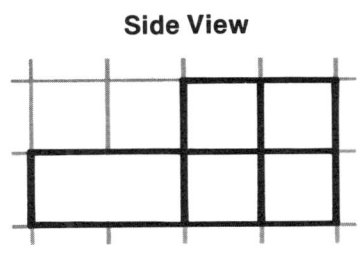

Side View

2) Now build this three-dimensional rod design using 1 purple, 2 light green, 2 red, and 2 white rods.

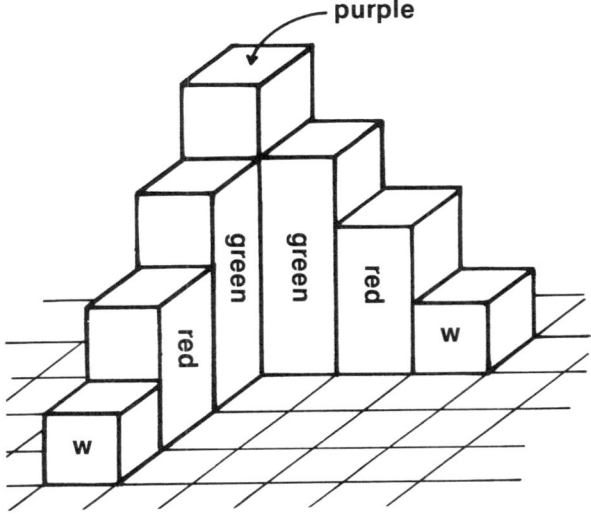

Draw the front view. The top and side views are given.

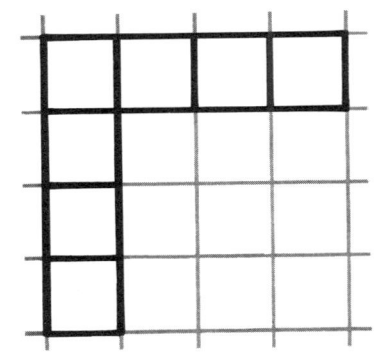

Top View

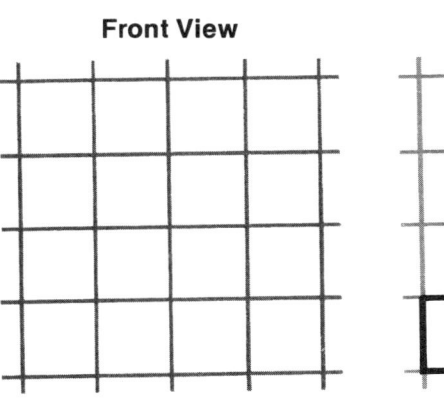

Front View

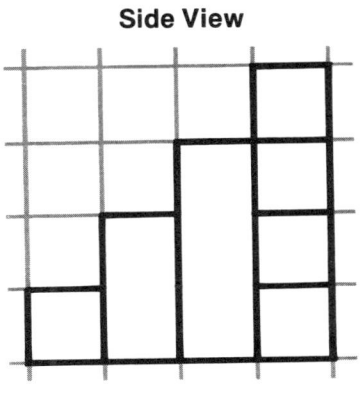

Side View

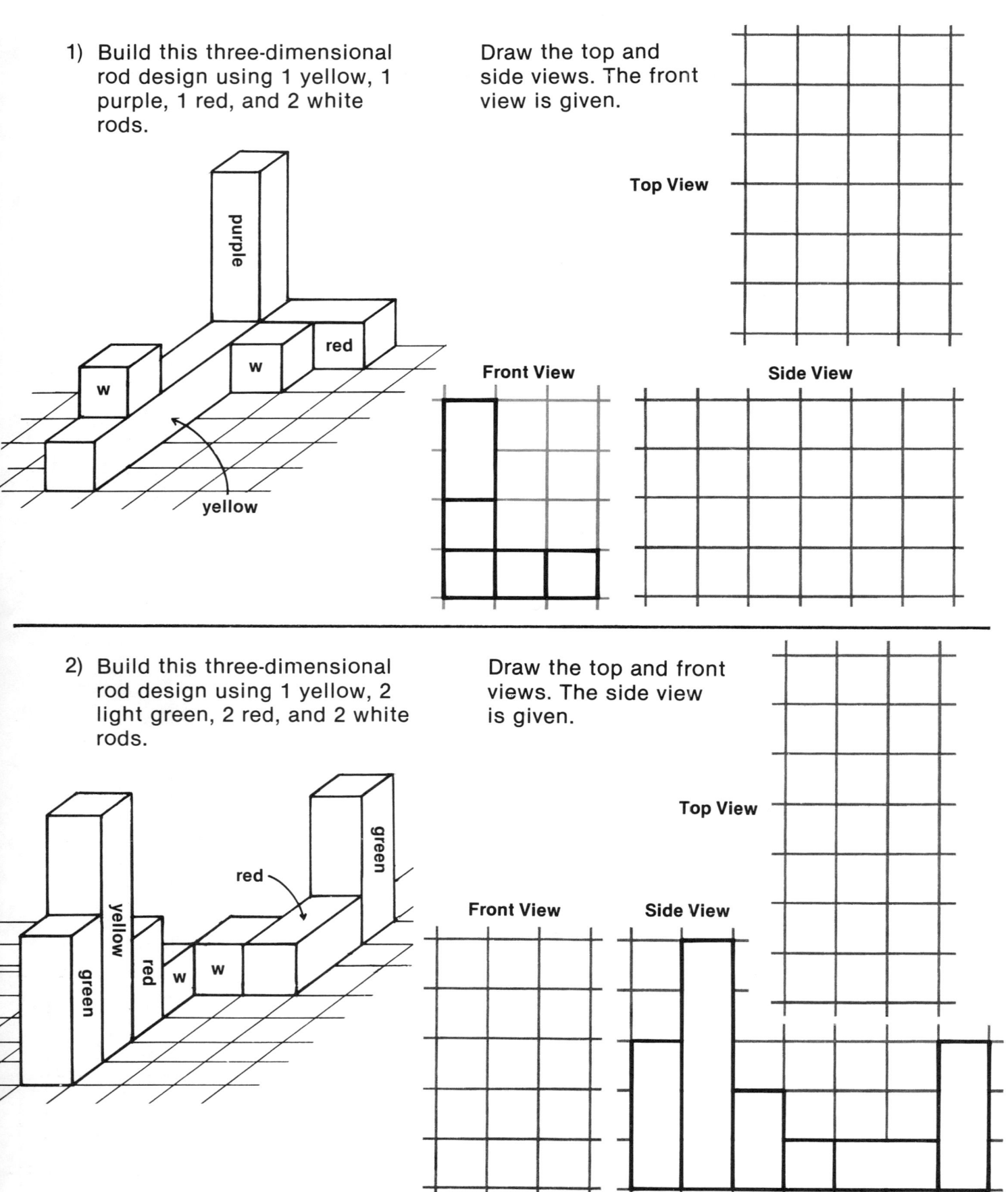

DRAWING THE THREE VIEWS OF A THREE-DIMENSIONAL ROD DESIGN

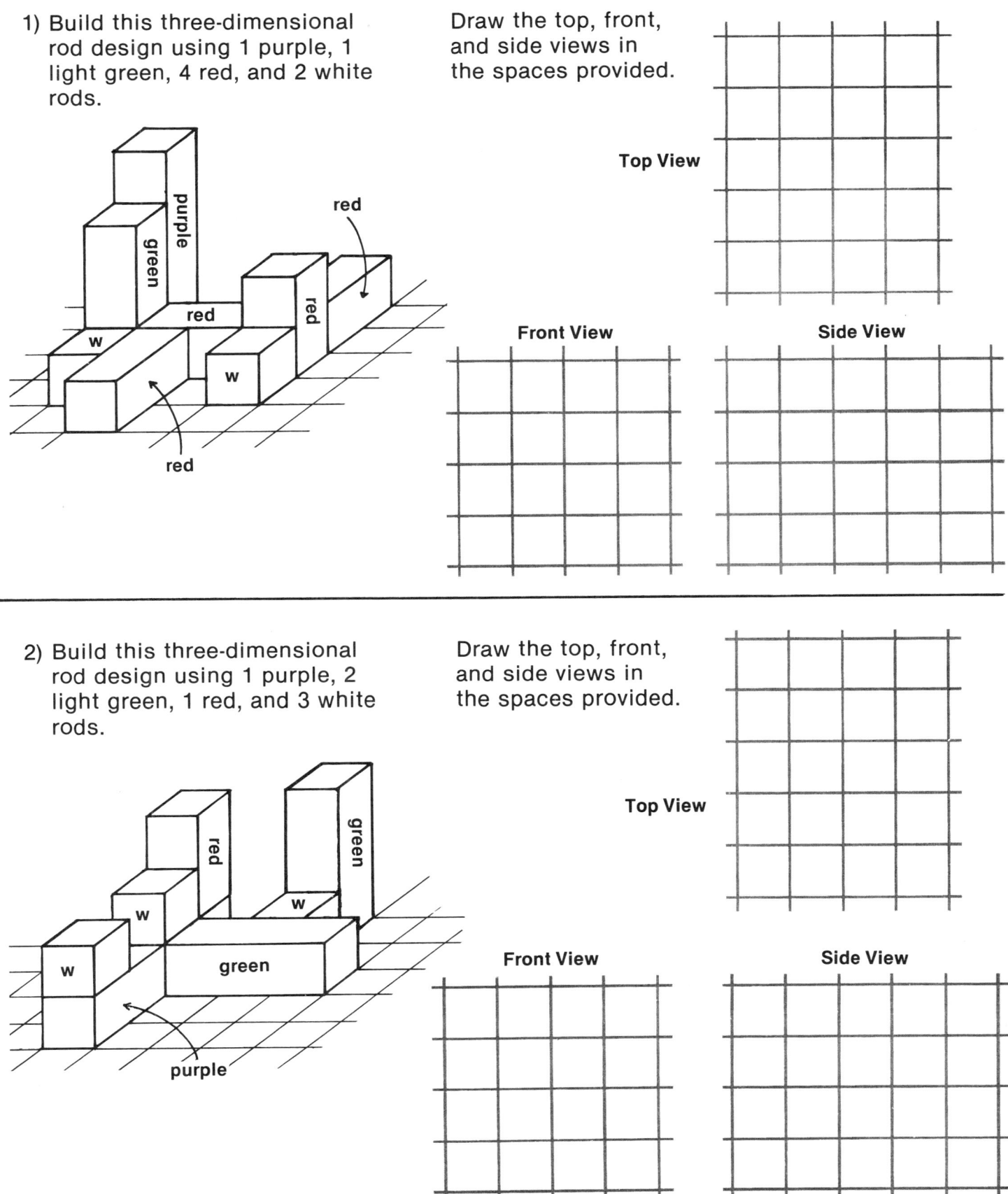

1) Build this three-dimensional rod design using 1 purple, 1 light green, 4 red, and 2 white rods.

Draw the top, front, and side views in the spaces provided.

Top View

Front View

Side View

2) Build this three-dimensional rod design using 1 purple, 2 light green, 1 red, and 3 white rods.

Draw the top, front, and side views in the spaces provided.

Top View

Front View

Side View

SPATIAL PROBLEM SOLVING with Cuisenaire Rods © 1983 Cuisenaire Co. of America, Inc.

DRAWING THE THREE VIEWS OF A THREE-DIMENSIONAL ROD DESIGN

1) Build this three-dimensional rod design using 1 purple, 2 light green, and 2 white rods.

Draw the top, front, and side views in the spaces provided.

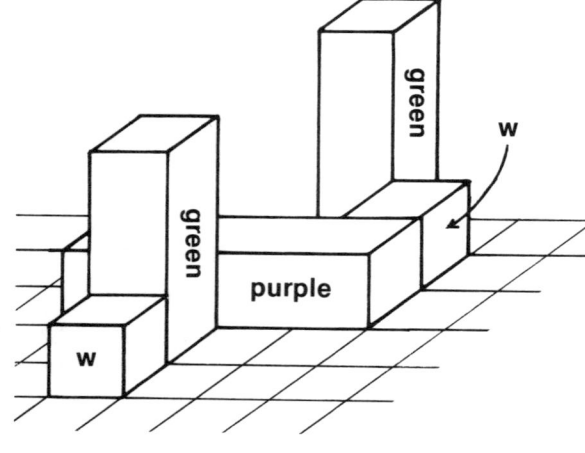

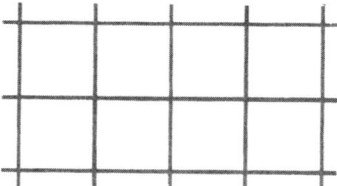

Top View

Front View

Side View

---

2) Build this three-dimensional rod design using 1 purple, 2 light green, 2 red, and 2 white rods.

Draw the top, front and side views in the spaces provided.

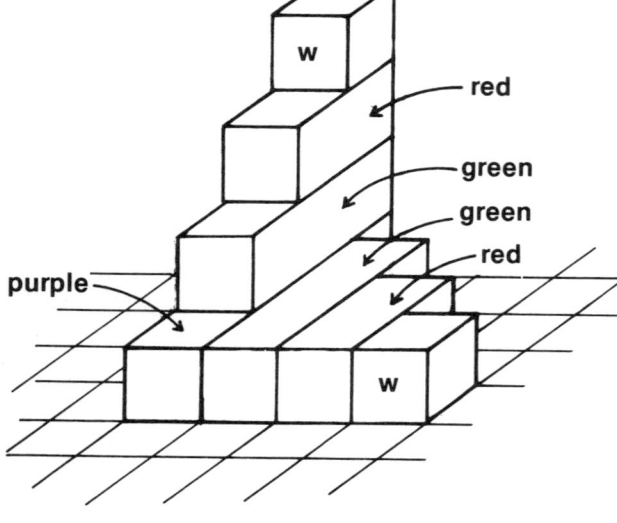

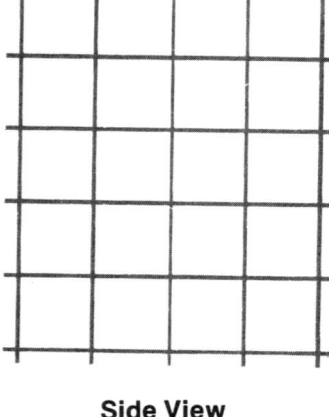

Top View

Front View

Side View

Page 45   Spatial Problem Solving with Cuisenaire Rods © 1983 Cuisenaire Co. of America, Inc.

DRAWING THE THREE VIEWS OF A THREE-DIMENSIONAL ROD DESIGN

1) Build this three-dimensional rod design using 1 yellow, 3 red, and 3 white rods.

   Draw the top, front, and side views in the spaces provided.

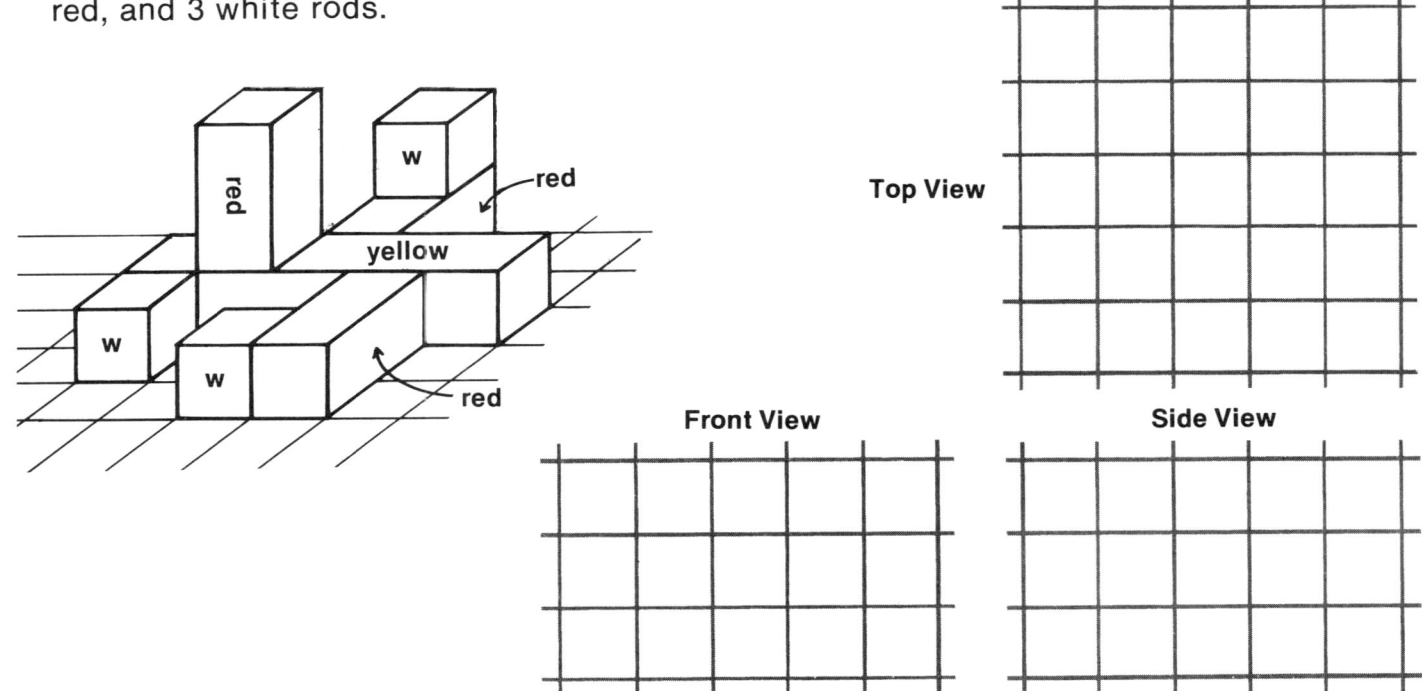

Top View

Front View

Side View

2) Build this three-dimensional rod design using 1 purple, 1 light green, 2 red, and 3 white rods.

   Draw the top, front and side views in the spaces provided.

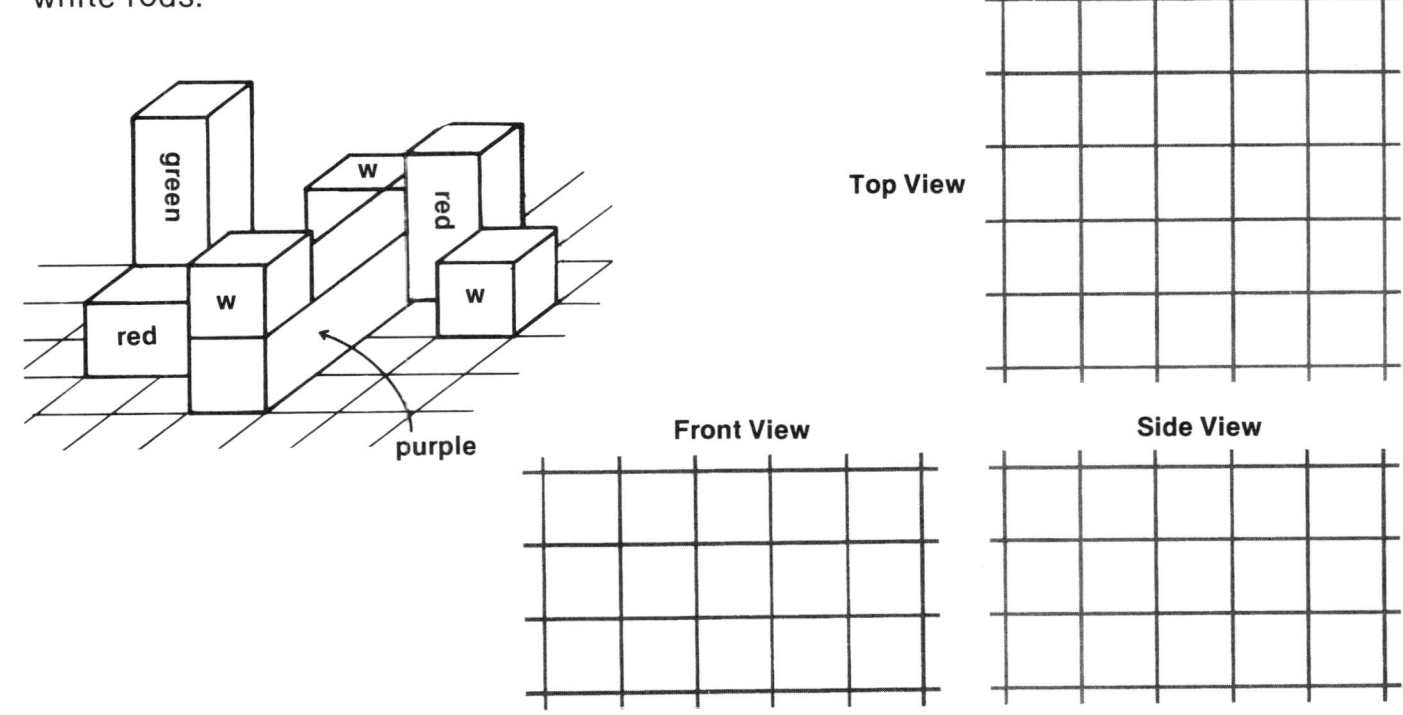

Top View

Front View

Side View

Spatial Problem Solving with Cuisenaire Rods © 1983 Cuisenaire Co. of America, Inc.

IDENTIFYING ROD DESIGNS FROM TWO-DIMENSIONAL DRAWINGS

Here are the top, front, and side views of two three-dimensional rod designs. Find which of the three rod designs at the bottom of the page goes with each set of drawings. Cross out the design that will not be used.

Set I:
Top View   Front View   Side View

Set II:
Top View   Front View   Side View

Rod Designs:

Which rod design matches the drawings in Set I? _____

Which rod design matches the drawings in Set II? _____

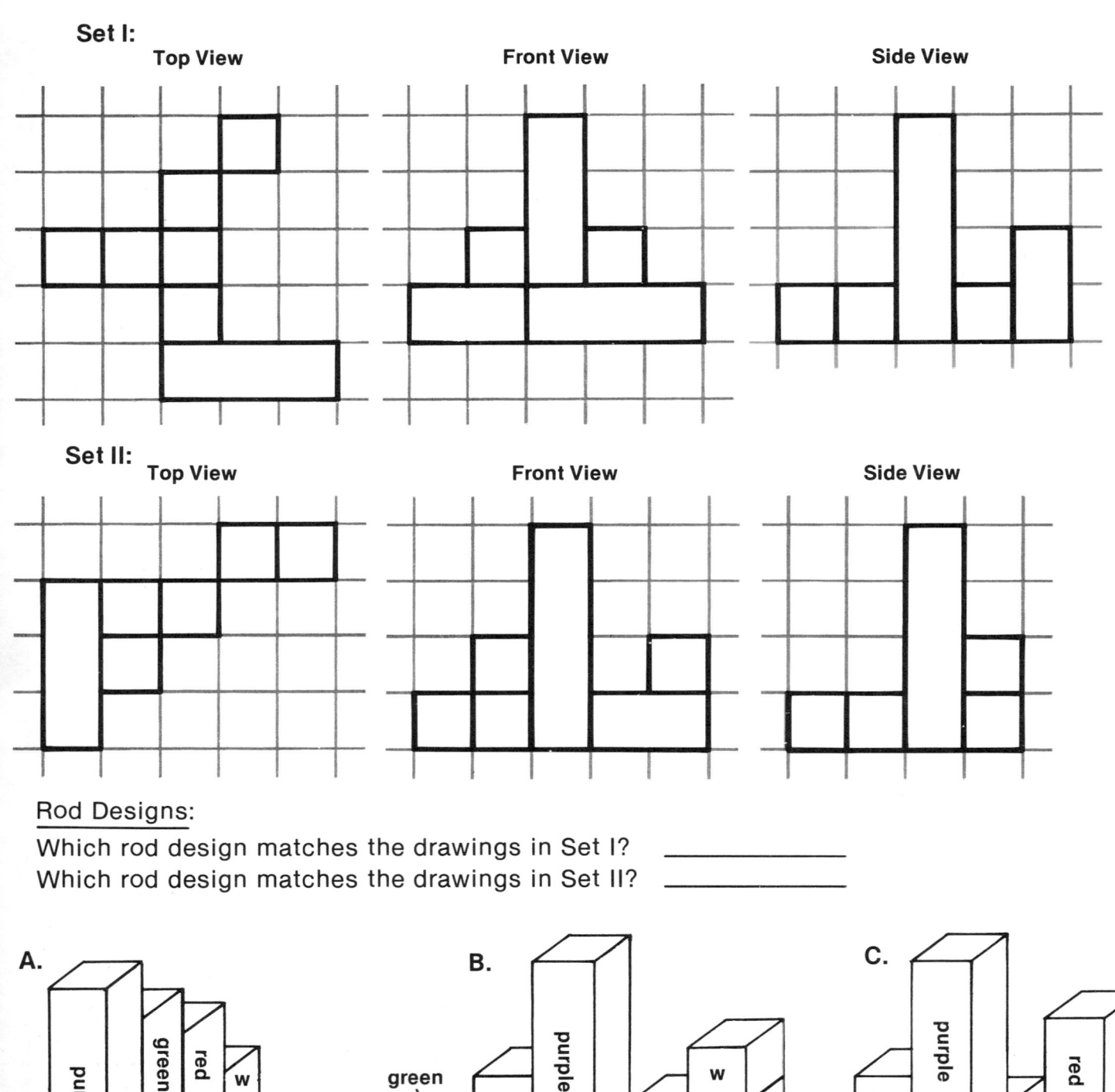

A. B. C.

Page 47

SPATIAL PROBLEM SOLVING with Cuisenaire Rods © 1983 Cuisenaire Co. of America, Inc.

IDENTIFYING ROD DESIGNS FROM TWO-DIMENSIONAL DRAWINGS

Here are the top, front, and side views of two three-dimensional rod designs. Find which of the three rod designs at the bottom of the page goes with each set of drawings. Cross out the design that will not be used.

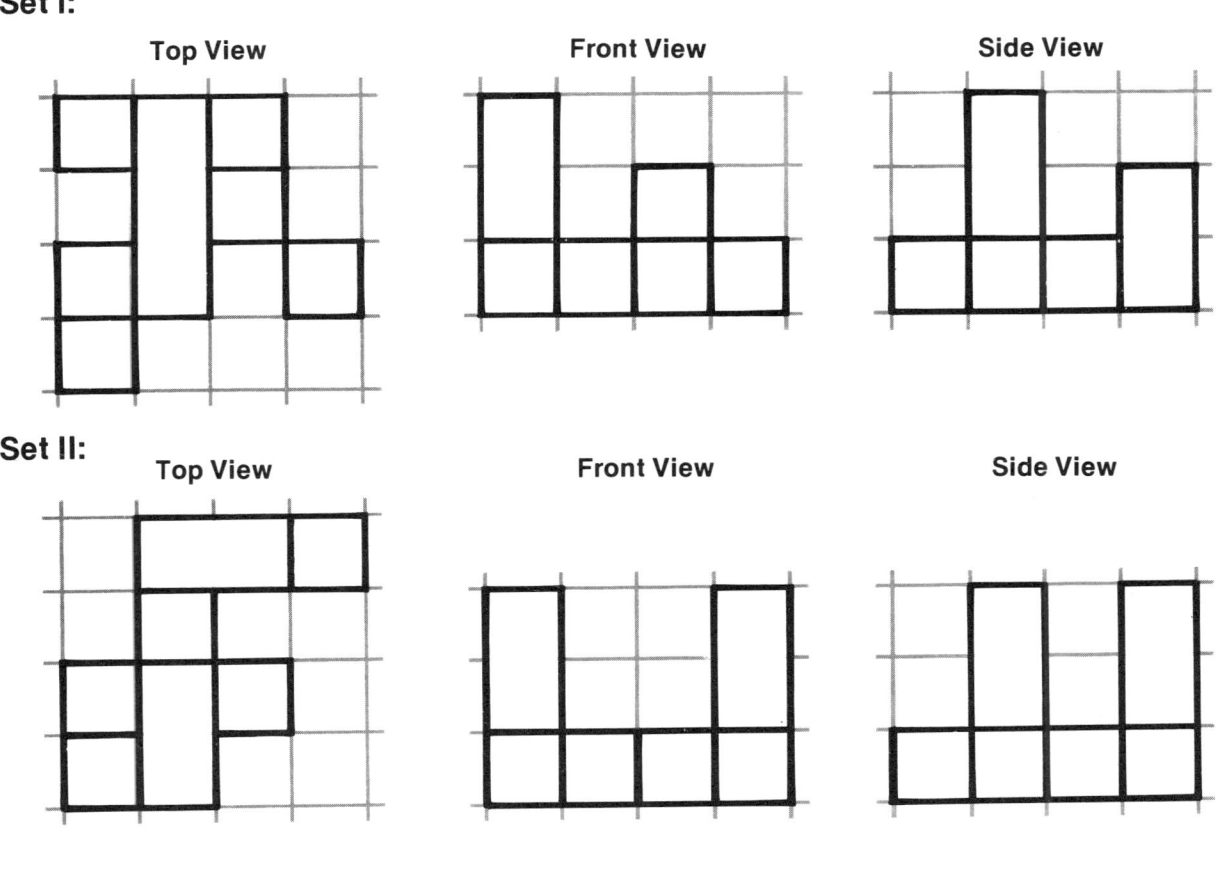

Rod Designs:

Which rod design matches the drawings in Set I? _____
Which rod design matches the drawings in Set II? _____

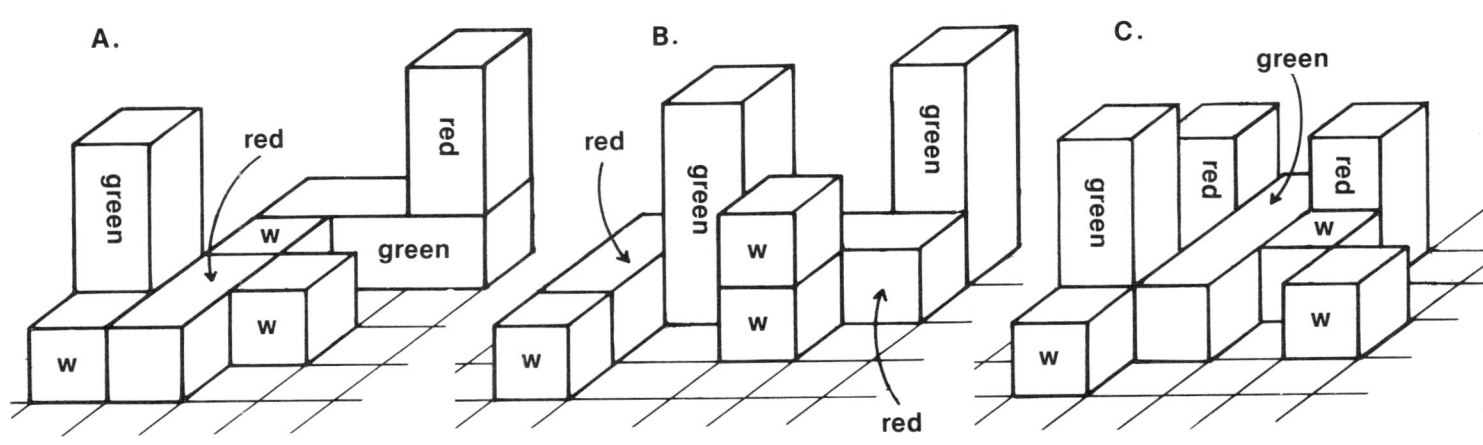

Spatial Problem Solving with Cuisenaire Rods © 1983 Cuisenaire Co. of America, Inc.

BUILDING ROD DESIGNS FROM TWO-DIMENSIONAL DRAWINGS

Here are the top, front, and side views of three-dimensional rod designs. Build the three-dimensional rod design that goes with each set of drawings. The rods you need are listed. They are the same for each set of drawings.

Set I: Use 1 yellow, 2 purple, 1 light green, 2 red, and 3 white rods.

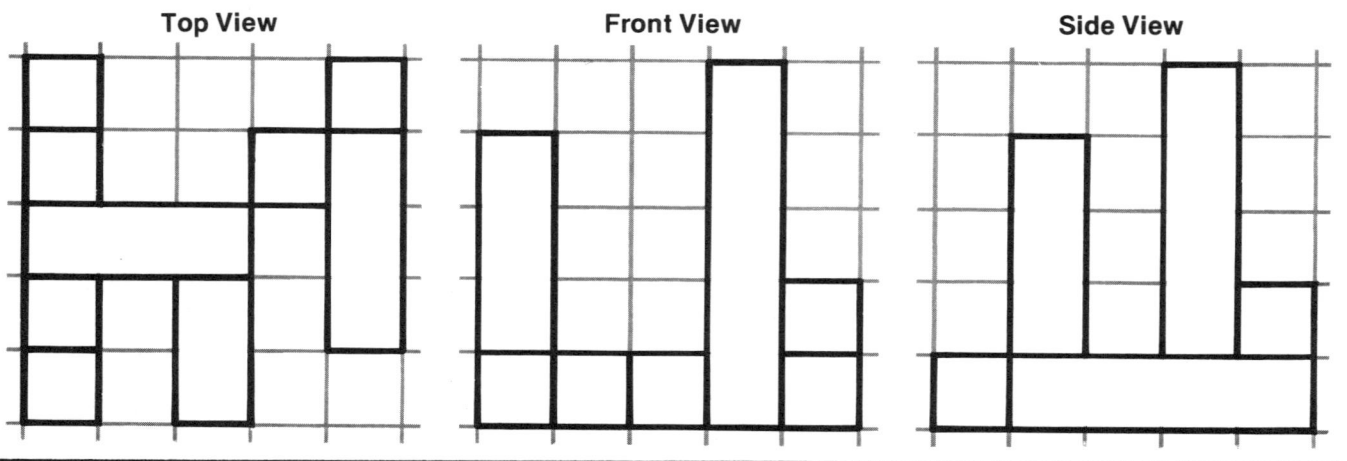

Set II: Use 1 yellow, 2 purple, 1 light green, 2 red, and 3 white rods.

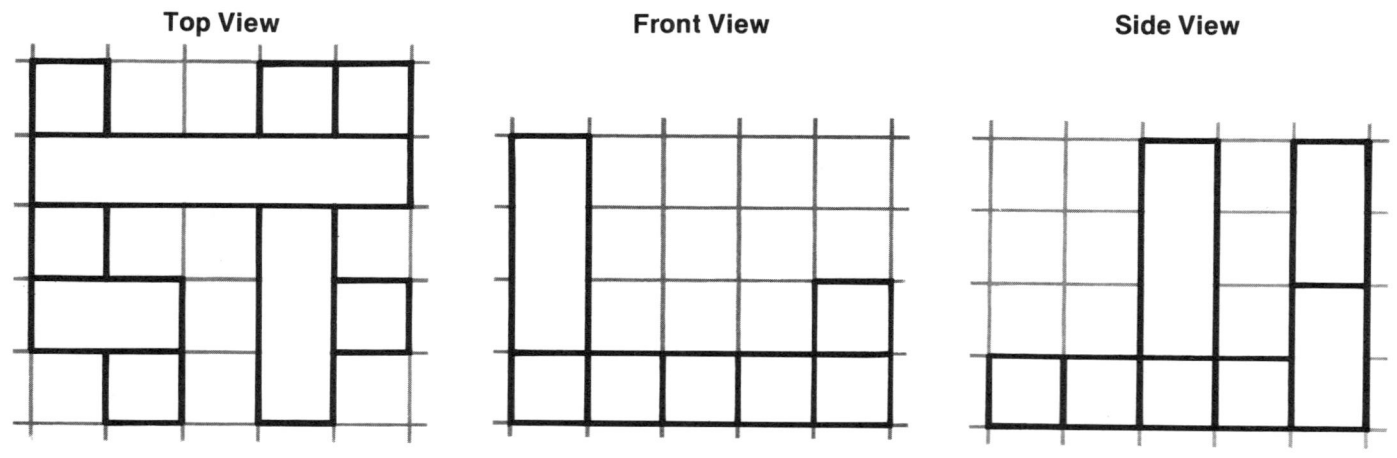

Set III: Use 1 yellow, 2 purple, 1 light green, 2 red, and 3 white rods.

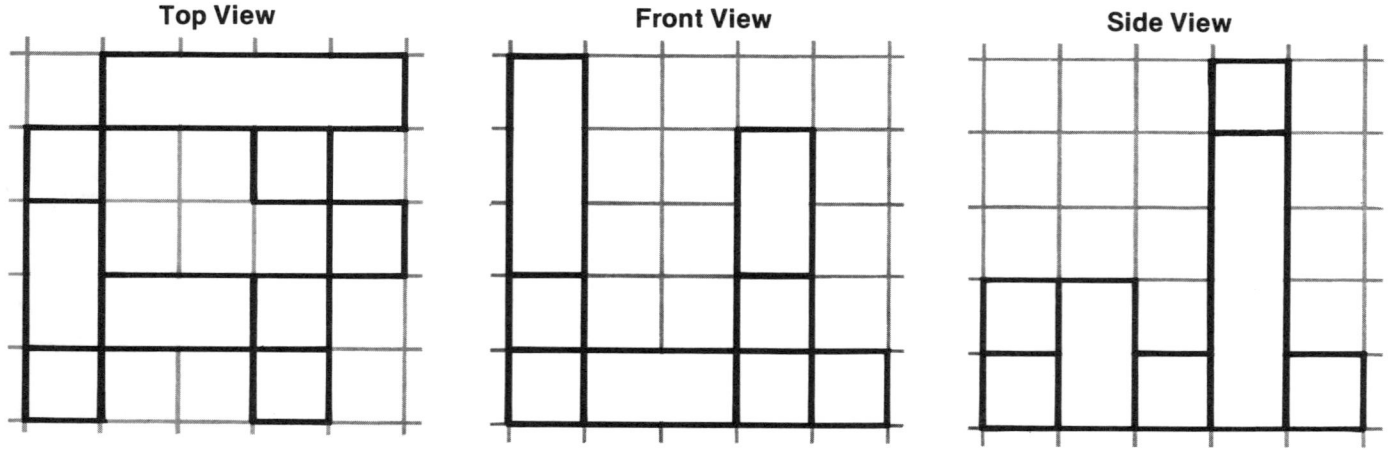

BUILDING ROD DESIGNS FROM TWO-DIMENSIONAL DRAWINGS

Here are the top, front, and side views of three-dimensional rod designs. Build the three-dimensional rod design that goes with each set of drawings. The rods you need are listed. They are the same for each set of drawings.

Set I: Use 1 yellow, 1 purple, 1 light green, 1 red, and 3 white rods.

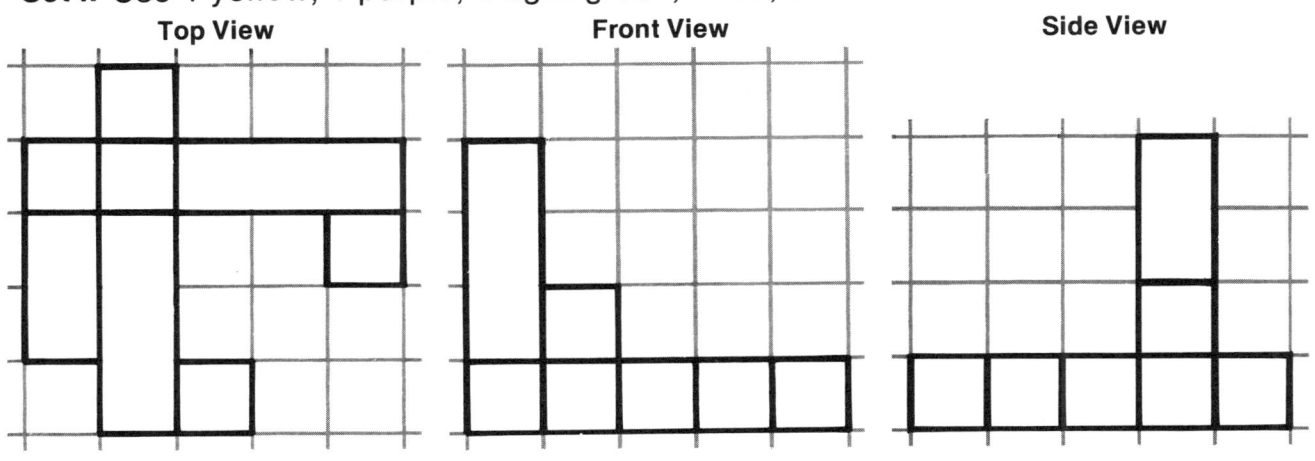

Set II: Use 1 yellow, 1 purple, 1 light green, 1 red, and 3 white rods.

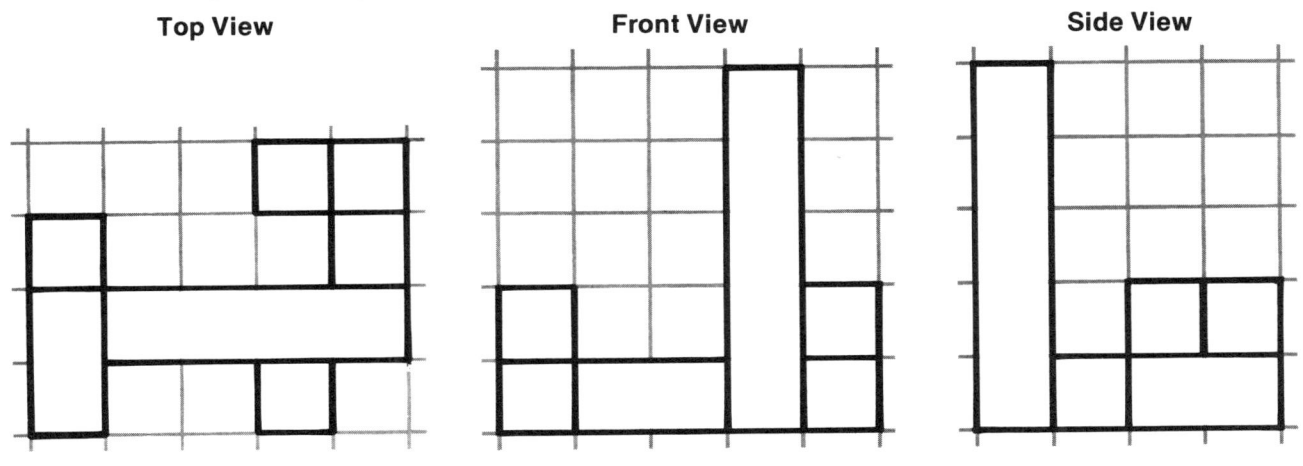

Set III: Use 1 yellow, 1 purple, 1 light green, 1 red, and 3 white rods.

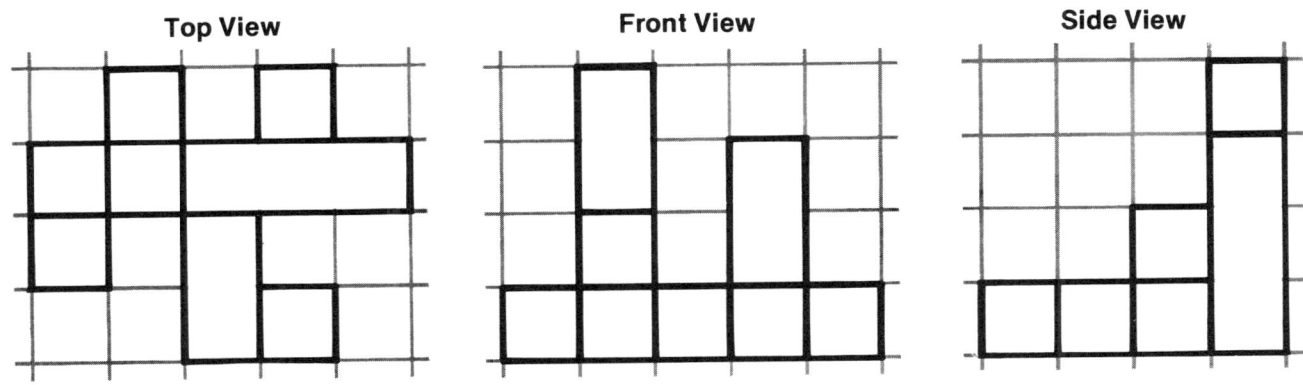

BUILDING ROD DESIGNS FROM TWO-DIMENSIONAL DRAWINGS

Here are the top, front, and side views of three-dimensional rod designs. Build the three-dimensional rod design that goes with each set of drawings. The rods you need are listed. They are the same for each set of drawings.

Set I: Use 1 yellow, 2 purple, 2 light green, 2 red, and 3 white rods.

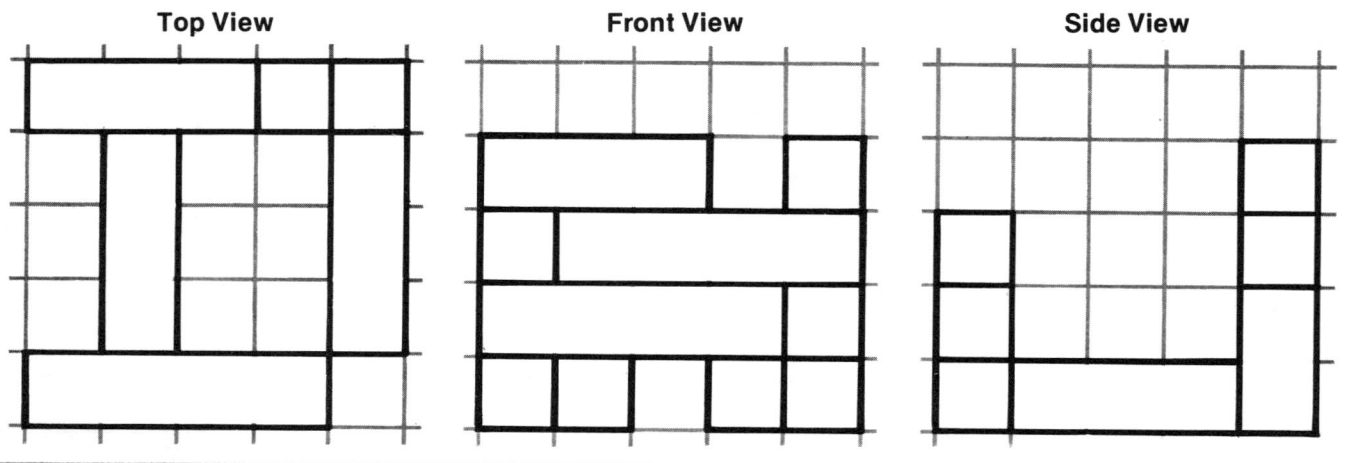

Set II: Use 1 yellow, 2 purple, 2 light green, 2 red, and 3 white rods.

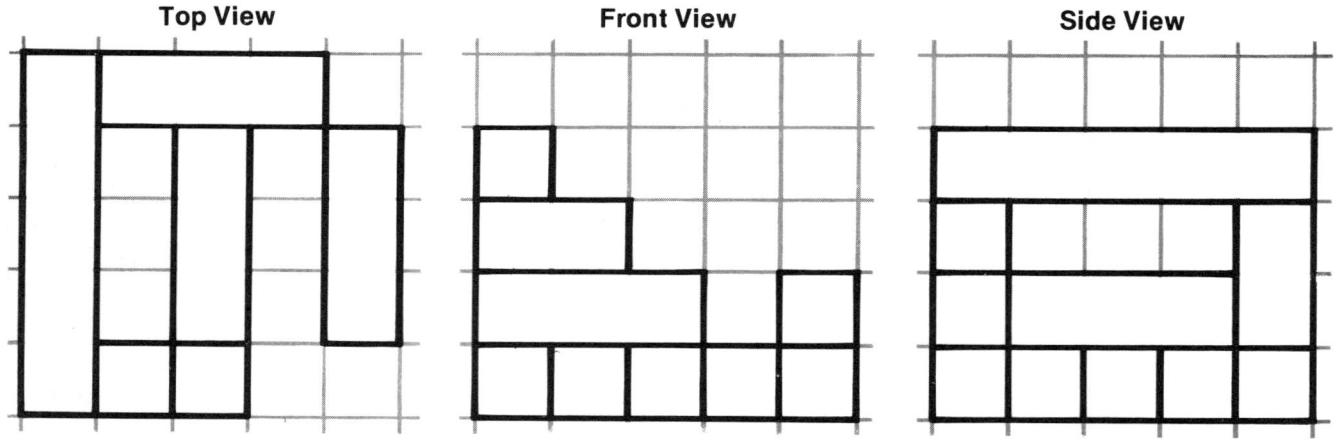

Set III: Use 1 yellow, 2 purple, 2 light green, 2 red, and 3 white rods.

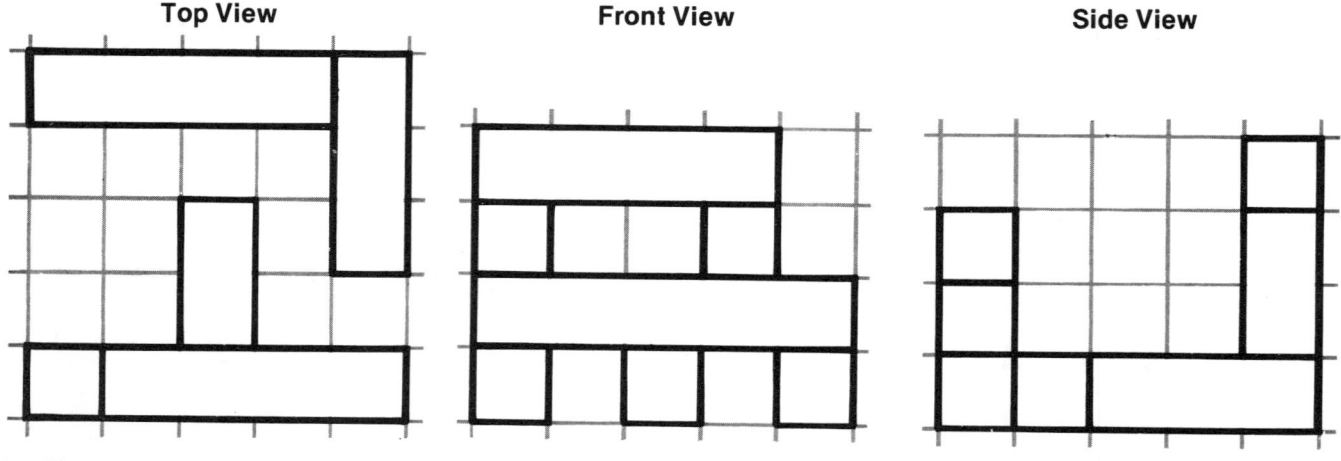

SELECTED ANSWERS AND COMMENTS

Pages 1-14: General Comments

These space-filling activities involve three types of problems: covering designs with one rod of each color, covering designs with a given set of rods, and covering designs with a specific number of rods. The third type, Pages 9-14, is the most challenging and should be preceded by ample experience with the first two types. However, the level of difficulty of particular designs within each type of problem may vary from individual to individual. Students should be given as much time as they need on each problem and should be encouraged to move on if a particular design gives them trouble. They can come back to that design at a later time. Some students may do these tangram-like activities with speed and ease; while others will require much time and many tries. Be prepared to find that the students who excel on these tasks may not be the same students who excel on computational tasks, and that some of your numerical whizzes may be stumped here. Once students discover the strategy to place the large rods first, the solutions become easier to obtain.

Some problems have only one solution; whereas others have more than one solution. The initial goal for these problems should be for students to find just one solution for each design. But as students share their work, other solutions will emerge. Many students will become motivated to find all possible solutions. A great deal of fruitful problem solving is involved in reasoning out why a particular design has a unique solution. Students should be encouraged to discuss their arguments with their classmates. There are so many levels on which these designs can be approached that they meet the needs of a variety of students and can be explored several times throughout the year.

Page 1:

Two possible solutions are shown which are quite different — in one case, the orange rod is placed horizontally; and in the other, vertically. This design has a great deal more flexibility than the one on page 2; hence students should be encouraged to find many ways of solving it.

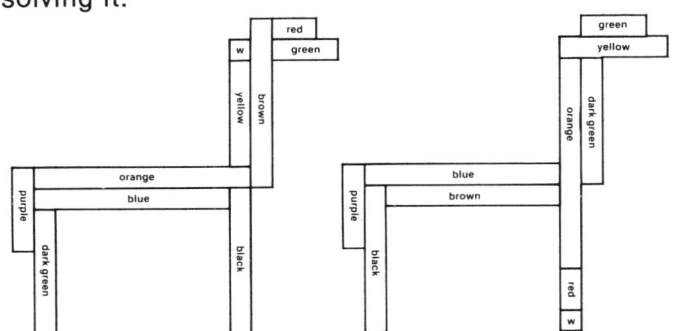

Page 2:

Two possible solutions are shown which are quite similar. In contrast to page 1, this design offers less opportunity to change rods from a horizontal to a vertical orientation. The number and variety of solutions are limited. If students have difficulty finding even one solution they should move on to page 3 which may seem easier to them and come back to this design at a later time.

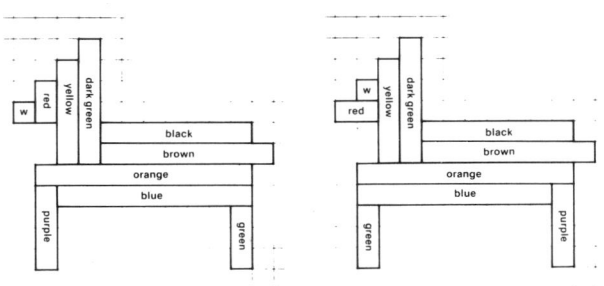

Page 3: 1)

This is an easy problem for most students since there are many solutions and a great deal of flexibility in the placement of the rods. Two possible solutions are shown.

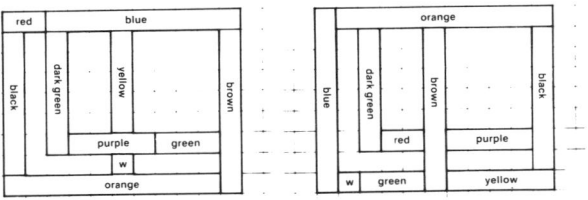

2)

In order to get a solution for this design, the longest rods (orange, blue, and brown) can be placed only in the locations shown, illustrating how crucial it is to place the big rods first. The small rods offer some flexibility.

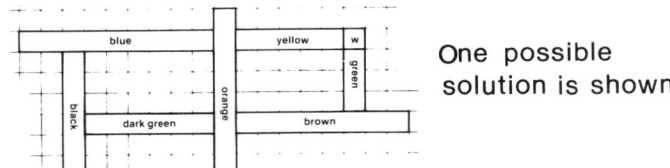

One possible solution is shown.

Page 4: 1)

This design has only one solution. Students should be encouraged to discuss why this is so. Such discussions enhance problem solving skills.

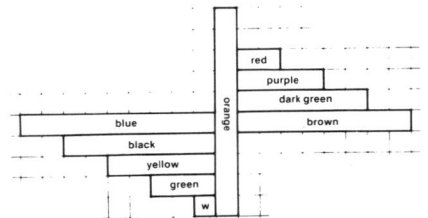

2)
This design can be done in more than one way. One possible solution is given.

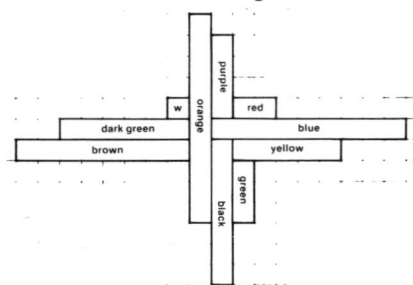

Page 5: 1)
This design can be done in more than one way. Two possible solutions are given.

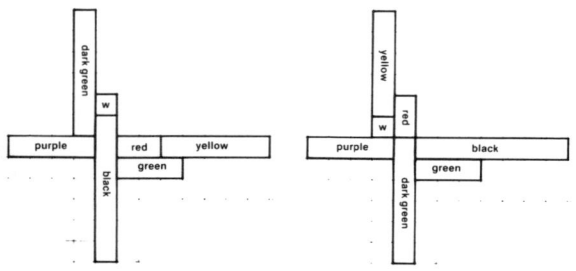

2)
This design has only one solution.

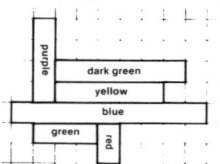

Page 6: 1)
This design can be done in more than one way. Two possible solutions are given.

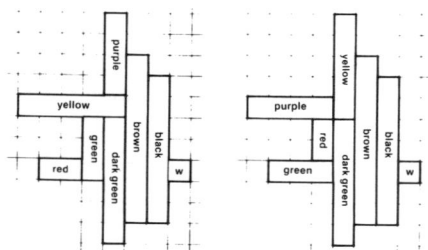

2)
This design can be done in more than one way. Two possible solutions are given. Note that the blue, black, and dark green rods have unique placements, but there is flexibility with the smaller rods.

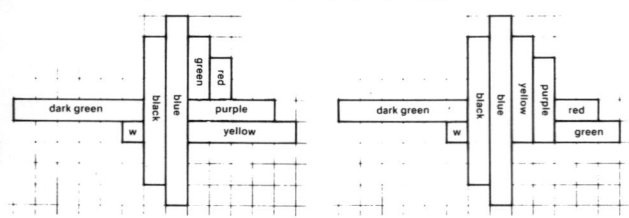

Page 7:
Both of these designs have a great deal of flexibility. Two possible solutions are given for each.

1)

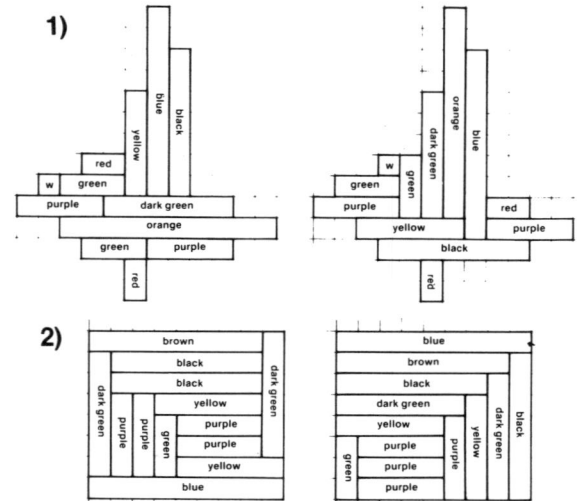

Page 8: Possible solutions are shown.

1)

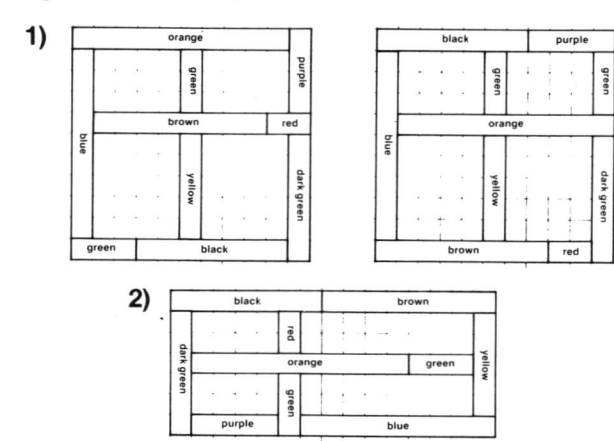

Page 9:
Students may wish to use numerical as well as spatial arguments as they solve this type of problem. They will find that some rods are "too large" and some are "too small".

The set of rods that can be used is unique, but there can be slight modifications in their placement. This solution in unique.

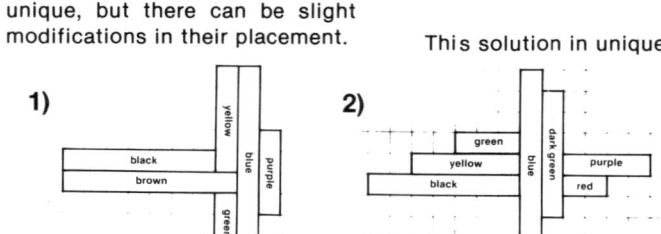

Page 10:
Two possible solutions are given.

1)

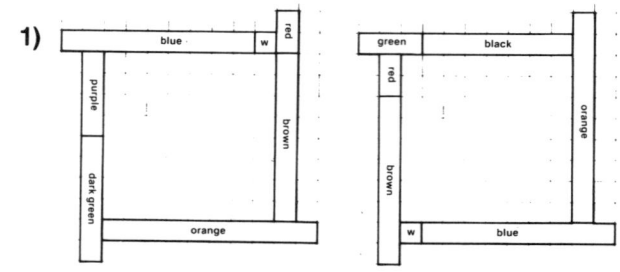

Page 53 SPATIAL Problem Solving with Cuisenaire Rods © 1983 Cuisenaire Co. of America, Inc.

Page 10: 2)
Two solutions are given. It should be noted that they involve the same choice of rods which are placed in different arrangements to produce the design. An interesting topic for discussion is why the choice of rods is unique even though their placement is not.

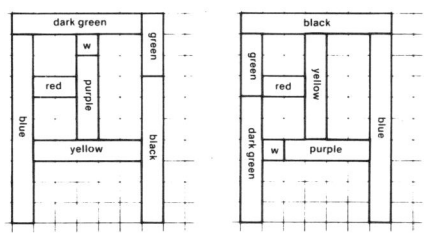

Page 11: 1)
The choice of rods is unique, but several arrangements are possible. Two possible solutions are shown.

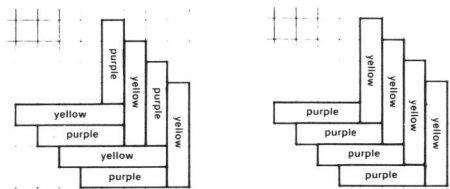

2)
Two different solutions with different sets of rods are possible.

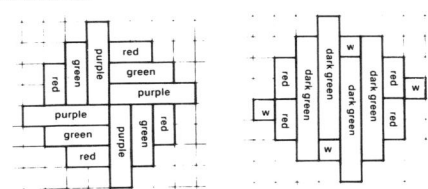

This solution in unique. This set of rods can be modified in their placement.

Page 12: 1)
Two possible solutions are given for each design.

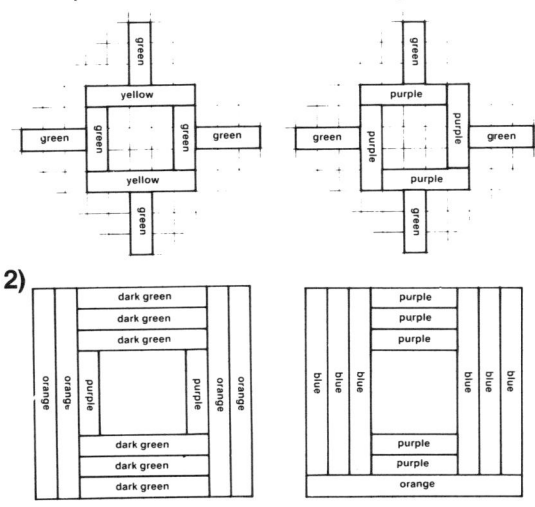

Page 13: 1)
The solutions to this design lend themselves to patterning. For example, there can be 4 of each of the three colors — red, green, and dark green, placed in the following configuration.

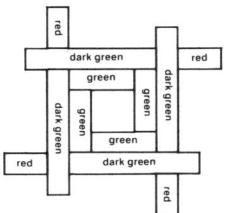

Another solution involves 2 dark green, 4 yellow, and 6 red rods, placed as follows. Again the patterns are interesting.

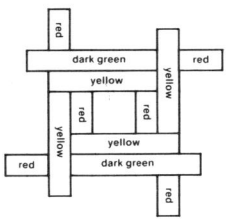

There are other solutions as well. No solution can use orange, blue, or brown rods. The students should argue why this is so.

2)
Students may find this problem difficult to solve, as the solution is unique. They may be able to reason that it is impossible to use orange, blue, or brown rods. Once these "long" rods are eliminated, "small" rods like white, red, and green also need to be eliminated. Even with these insights, the choice and placements of the rods remain a challenge.

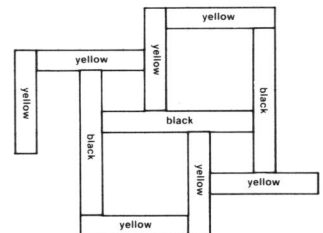

Page 14: 1)
The solution to this design is unique. Students can argue that it is impossible to use orange, blue, brown, or black.

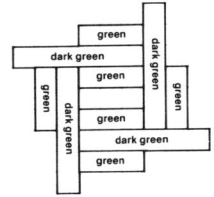

A different problem may be posed for this same design. "Use exactly twelve rods and only two colors". The solution this time is 6 yellows and 6 reds.

2)
Two possible solutions are given to what may seem like another challenging problem because of the restrictions. Students should be encouraged to reason out what rods are impossible to use and to find other solutions besides these two.

Spatial Problem Solving with Cuisenaire Rods © 1983 Cuisenaire Co. of America, Inc.

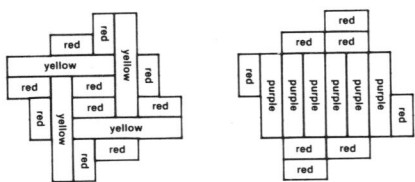

Pages 15-20: General Comments

These six pages give students practice with three types of rotations and are prerequisite (along with pages 21-26 on reflections) to pages 27-40 which combine the two motions. Although the space filling activities on Pages 1-14 need not precede pages 15-20, most students will start with them.

The idea of motion in geometry is an important one; for example, children are frequently observing the wheels of a bike or the hands of a clock. This section only deals with three types of turns in a clockwise direction: 90° or a quarter turn, 180° or a half turn, and 270° or a three-quarter turn. Arrow notation is used to help students remember the motion to be done. Using rod designs, students can visualize the motion that would transform an original rod design into the resulting design. However to carry out the motion physically is difficult since the rods may slip and slide. Gluing together the original rod designs remedies this problem and should be done for demonstration purposes and for individuals who need more concrete experience. It should be noted that most glues hold rods together permanently; but that rubber cement, especially if applied to both surfaces to be fastened, provides a temporary structure so that the rods can later be recycled back into normal use.

All students can use a cut-out of the rod design to rotate physically according to the prescribed directions. Teachers should duplicate a copy of the Master Cut-out Sheet for each student. Students should cut out the appropriate design as needed for a particular page.

Page 15:
This page requires no written work. The goal is to teach students three types of clockwise rotations: 90° ↘, 180° ↰, and 270° ↰. Students may find it physically difficult to rotate the actual rod designs (unless they glue the rods together). Hence the Master Cut-out #1 (on page 60) is provided. The picture of the design should be cut out and rotated in the prescribed ways. (Some students may observe that a 270° turn clockwise has the same result as if a 90° counterclockwise turn were done instead.)

Page 16:
Students should use a copy of the Master Cut-out #2 to perform these rotations. The rotated designs look like:

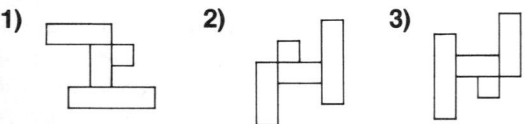

Page 17:
Students are asked to rotate four different designs. Some students will already be able to visualize the resulting designs; others may wish to make their own cut-outs by drawing the original designs on centimeter graph paper and cutting them out. Students should then physically do the rotations with the cut-outs.

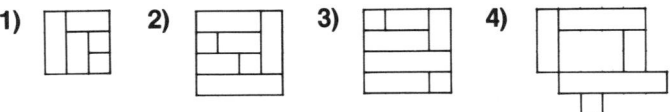

Page 18:
Students are asked to rotate three different designs. If they cannot visualize the resulting designs, they should make their own cut-outs by drawing the original designs on centimeter graph paper and cutting them out. Students should then physically do the rotations with the cut-outs.

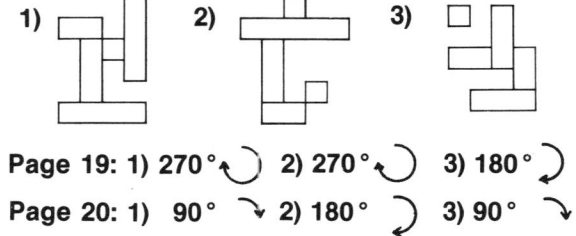

Page 19: 1) 270° ↰ 2) 270° ↰ 3) 180° ↰
Page 20: 1) 90° ↘ 2) 180° ↰ 3) 90° ↘

Pages 21-26: General Comments

These six pages give students practice with four types of reflections and are prerequisite (along with pages 15-20 on rotations) to pages 27-40 which combine the two motions.

Students may already be familiar with the notion of reflection from such activities as looking into a pool of water or paper folding. The four motions introduced here are: Vertical reflection, V ┆; Horizontal reflection, H ----; Left Diagonal reflection, L ╲; and Right Diagonal reflection, R ╱. The notation helps students remember the motion to be done.

To carry out the reflections with actual rod designs is physically difficult, as the designs fall apart in the attempt to flip them over. Gluing together the original rod designs remedies this problem and should be done for demonstration purposes and for individuals who need more concrete experience.

All students should be provided with a cut-out of the rod designs so as to be able to carry out physically the reflections. Teachers should duplicate a copy of the Master Cut-out Sheet for each student. Students should cut out the appropriate designs as needed for a particular page. The outlines of the rods must be traced on the back side of each cut-out used in a reflection since the design gets "flipped over". Students should hold the cut-outs up to the light to see where to draw the lines.

Also students may find it useful to use a mirror placed at the correct angle (vertically, horizontally, or diagonally) to see a picture of the reflected design.

Page 21:
This page requires no written work. The goal is to teach students four types of reflection: vertical, V⋮ ; horizontal, H ---; left diagonal, L ╲ ; and right diagonal, R ╱. Students will find it physically difficult to reflect the actual rod designs (unless they glue the rods together). Hence a Master Cut-out #3 is provided. The picture of the design should be cut out. For the reflection cut-outs, it is necessary to trace the outlines of the rods on the back side of the cut-out. The best way to do this is to hold the cut-out up to the light. Students should physically do the reflections with the cut-out.

Page 22:
Students should use a copy of the Master Cut-out #4 to perform these reflections.

Page 23:
Students should use a copy of the Master Cut-out #5 to perform these reflections.

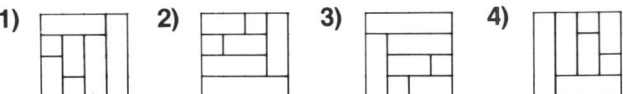

Page 24:
Students are asked to reflect four different designs. Some students will already be able to visualize the resulting designs; others may wish to make their own cut-outs using the centimeter graph paper on page 59.

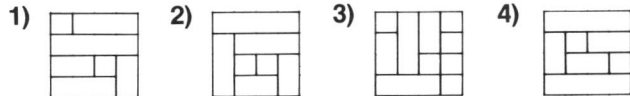

Page 25:
1) V ⋮ 2) R ╱ 3) L ╲ 4) H -----

Page 26:
1) H --- 2) V ⋮ 3) R ╱

Pages 27-40: General Comments
The activities in this section combine and apply the concepts of rotations and reflections which were introduced in the previous two sections. It should be noted that pages 15-26 are prerequisite to these pages which include analogies and chains of motions.

Students may already be familiar with the notion of an analogy. The interpretation here is based on rod designs related either by one of the types of rotations or one of the types of reflections:

"The first design is to its resulting design
as the second design is to what?"

Pages 27-40 give a wide variety of experience with analogies. Pages 35 and 36 ask students to reverse the process and find the second design, given the resulting design.

The chains are simply a sequence of motions devised so that if done correctly, they end with the original design. Master Cut-outs are provided to help students physically carry out the motions. The self-checking nature of these puzzles motivates students to "stay with" these problems even though they require several steps.

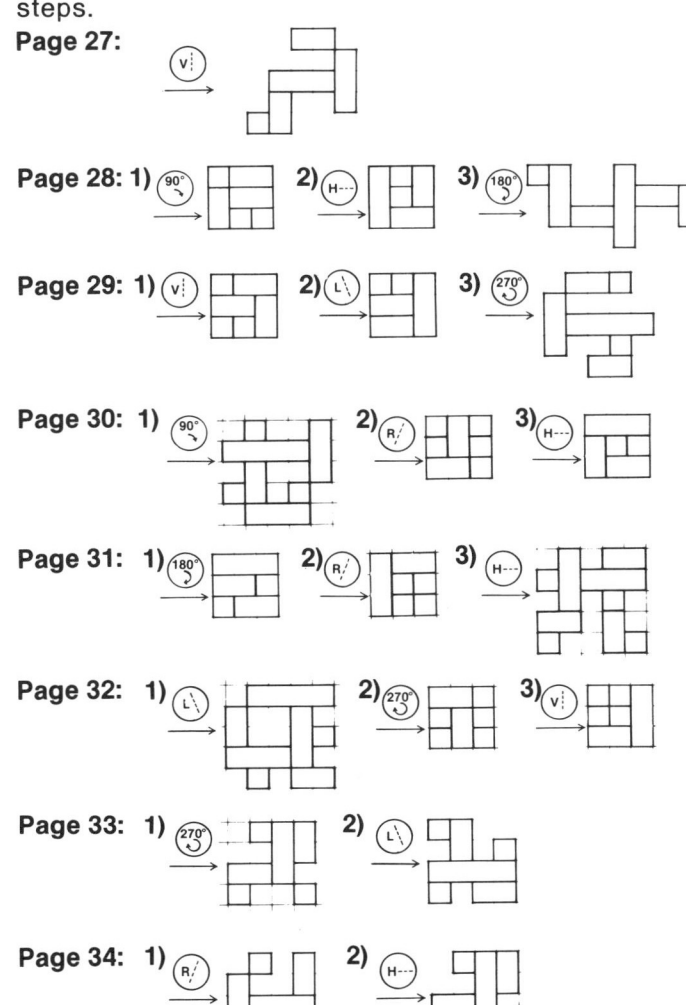

Page 35:
The missing designs this time are on the left side of the arrows. These analogy puzzles require reverse thought processes. "What second design would produce the final design?"

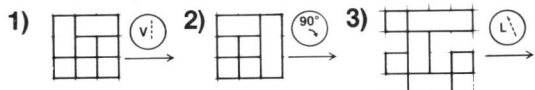

Page 36:
As with page 35, the missing designs are on the left side of the arrows.

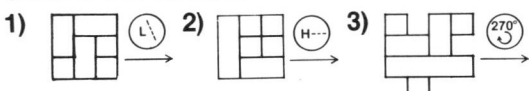

Page 37:
Students may need to try this chain more than once before they get it right; hence they should use a pencil rather than a pen to record their work at each stage. Some students may be able to visualize the rotations and reflections, but most students would benefit from

using a copy of the Master Cut-out #6. Since reflections are involved, it will be necessary for students to trace the outline of the rods on the back side of the cut-out. The best way to do this is to hold the cut-out up to the light. Students should physically do the rotations and reflections with the cut-out. The motions are done in succession. Students should start and end with the original design at the top of the page.

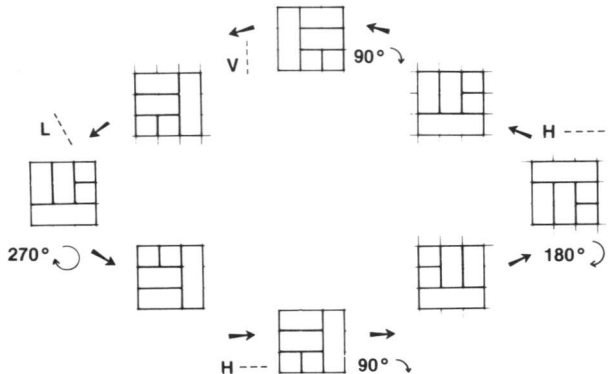

Page 38:
Students may use Master Cut-out #7 to help them do the rotations and reflections.

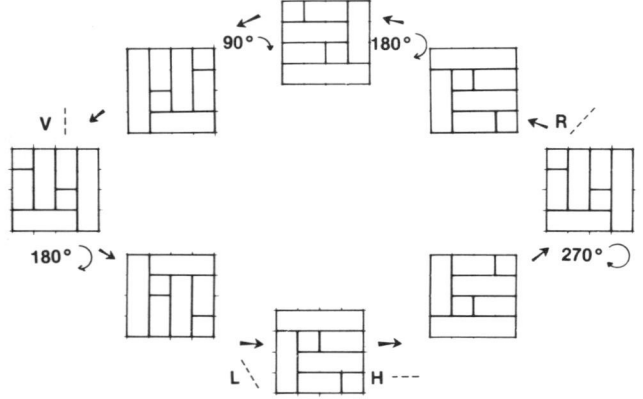

Page 39:
Students may use Master Cut-out #8 to help them do the rotations and reflections.

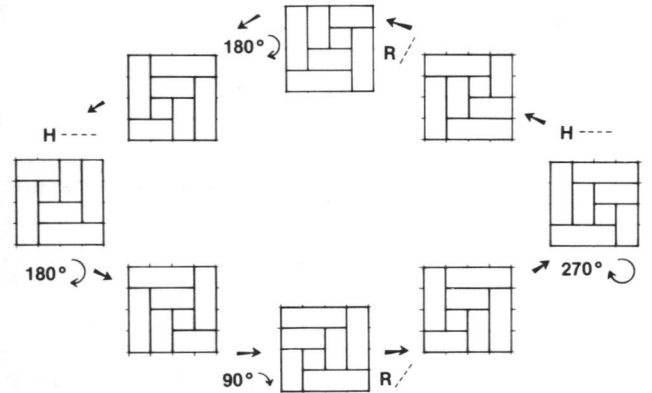

Page 40:
Students may use Master Cut-out #9 to help them do the rotations and reflections.

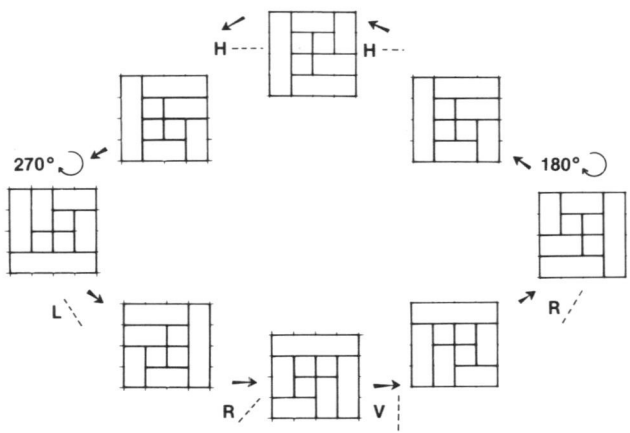

Pages 41-51: General Comments
This section stands alone and can be done at any time. It involves the relationship between three-dimensional designs and the two-dimensional views which are commonly drawn: top view, front view, and side view. It also asks students to reverse the process and find the three-dimensional design defined by the given two-dimensional views. In this section, students can fantasize being future architects as they get a first-hand experience with these important spatial skills.

Pages 41-43 are carefully sequenced to carry along the "beginner". As with the Tangram-type activities on Pages 1-14, specific suggestions are made for doing easier and harder versions of the same task. Pages 49-51 are difficult and open-ended and should be used to challenge and motivate only the exceptionally strong spatial problem solvers.

Page 41:
This page requires no written work. The goal is to teach students how to draw the two-dimensional top view, front view, and side view of a given three-dimensional rod design. Students should build the designs with rods and observe the design from the three viewpoints to see how the drawings are done.

Page 42:
Only one view for each design is required. Students should observe how the other two views are drawn in each case.

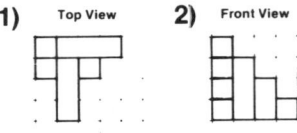

Page 43:
The students are given only one of the views and are asked to draw the other two views for each design.

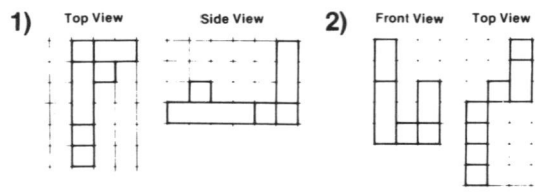

Page 44:
The students are asked to draw all three views.

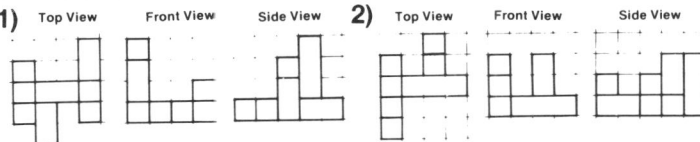

Page 45:

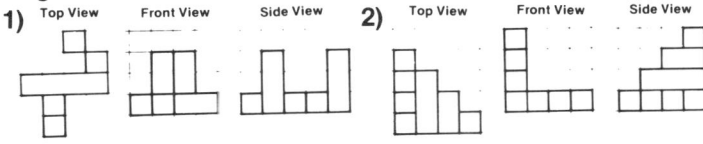

Page 46:

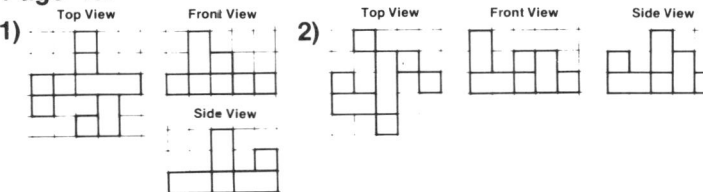

Page 47:
This matching task is a preparation for students to be able to build the three-dimensional design for a given set of two-dimensional top, front, and side views.
Set I: C   Set II: B

Page 48:
As with Page 47, this matching task is a preparation for students to go from two-dimensional drawings to the three-dimensional design they would generate.
Set I: C   Set II: A

Page 49:
This page has the more difficult task of building the three-dimensional design from the two-dimensional views.

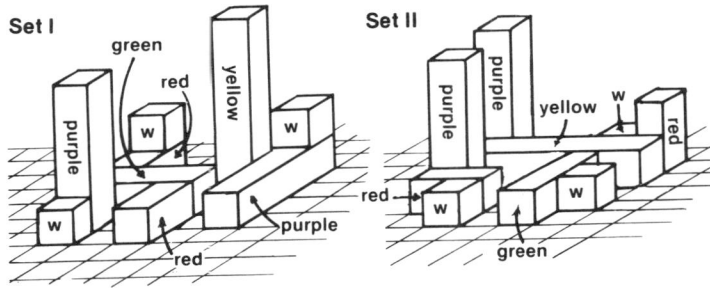

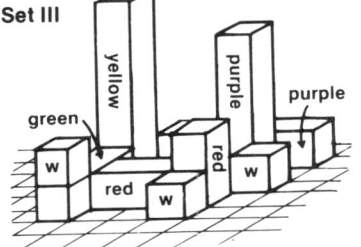

Page 50:
This page also poses the more difficult task of building a three-dimensional design from the two-dimensional views. Students should build each three-dimensional design. Then they should check their answer by drawing the three views for each on centimeter graph paper using the Graph Paper Master provided on page 59.

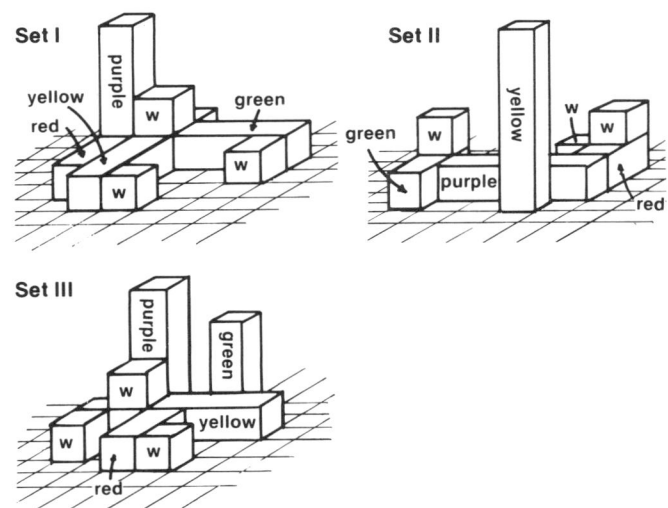

Page 51:
These three rod designs are very challenging since they have open spaces. Only the most spatially adept students will be able to do these. These students may even want to take this activity further by pretending that they are futuristic architectural engineers who like to work with open spaces. They will then make their own three-dimensional rod designs as models of buildings of the future. Then on graph paper, they should draw and label the three views for each design and present them as architectural plans.

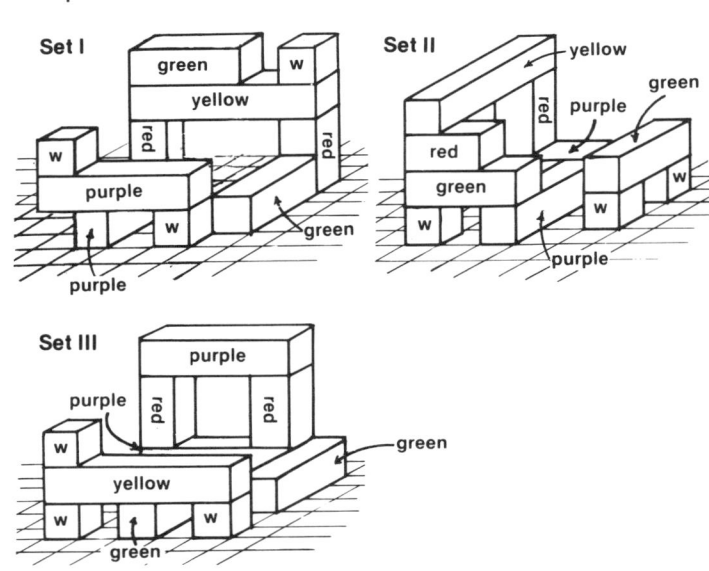

CENTIMETER GRAPH PAPER MASTER

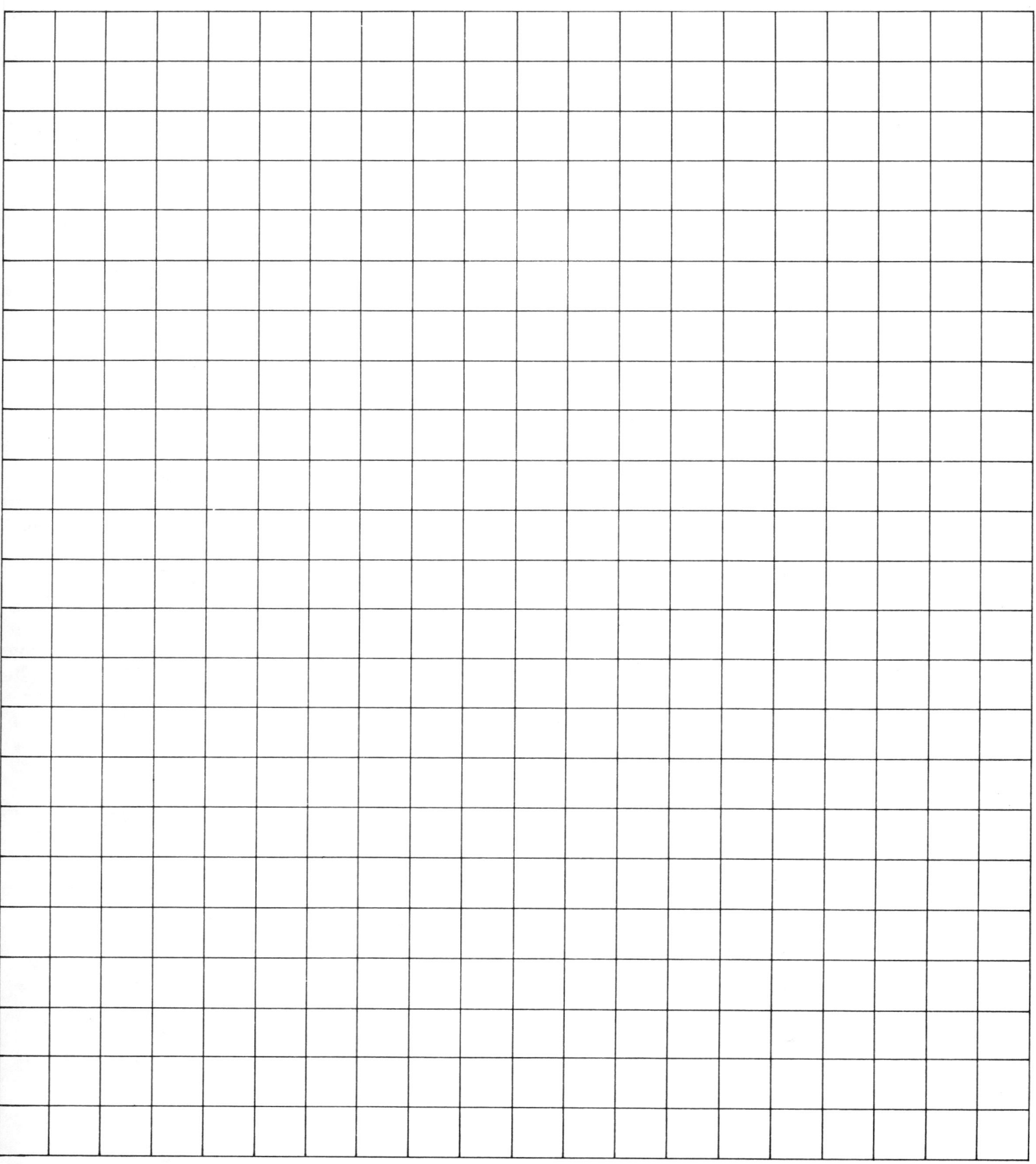

MASTER CUT-OUT SHEET

These rod designs may be cut out to be rotated or reflected as designated on each of the pages lised below. Specific instructions for their use are given in the Selected Answers and Comments Section for each of the pages.

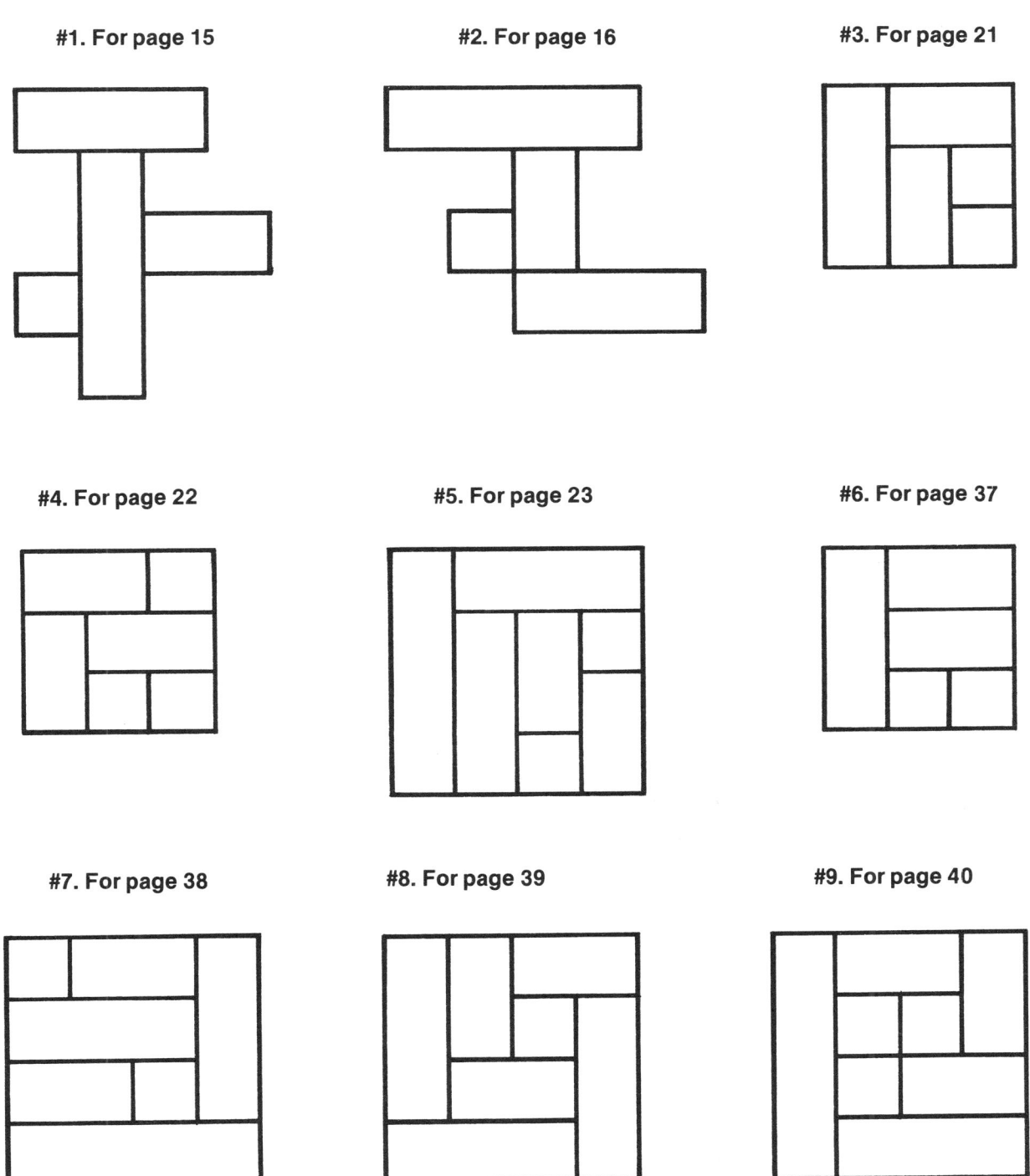

#1. For page 15

#2. For page 16

#3. For page 21

#4. For page 22

#5. For page 23

#6. For page 37

#7. For page 38

#8. For page 39

#9. For page 40

SPATIAL Problem Solving with Cuisenaire Rods © 1983 Cuisenaire Co. of America, Inc.